New Lands and Old Traditions

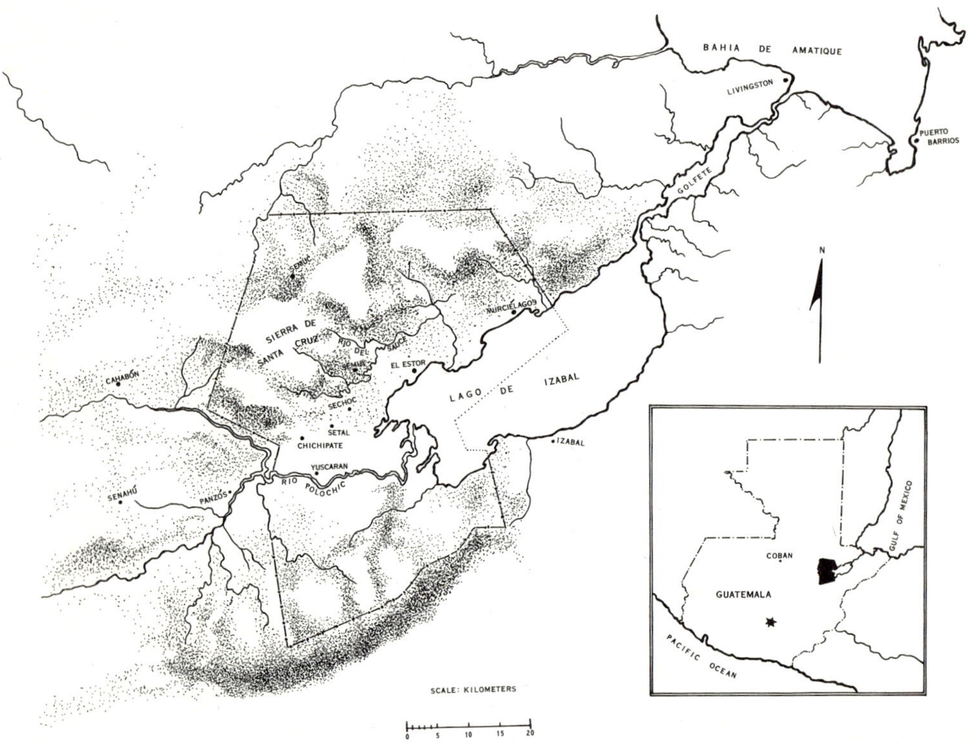

Polochic River, Lake Izabal, and the *municipio* of El Estor (only selected settlements are shown).

NEW LANDS and OLD TRADITIONS
Kekchi Cultivators in the Guatemalan Lowlands

WILLIAM E. CARTER

With the assistance of

SAMUEL C. SNEDAKER

UNIVERSITY OF FLORIDA PRESS

GAINESVILLE/1969

LATIN AMERICAN MONOGRAPHS—SECOND SERIES
NUMBER 6

A University of Florida Press Publication
SPONSORED BY THE
CENTER FOR LATIN AMERICAN STUDIES

LIBRARY OF CONGRESS
CATALOG CARD NO. 75-625422
SBN 8130-0267-2

PRINTED FOR THE PUBLISHER BY CONVENTION PRESS
JACKSONVILLE, FLORIDA

Preface

FOR SOME TEN YEARS various individuals from the University of Florida's College of Agriculture have been involved in research on the *municipio* of El Estor in the Izabal region of northern Guatemala. Stimulated by the original interest of Hugh Popenoe, a soils scientist, they have done work in general ecology, palynology, meteorology, archaeology, zoology, botany, and geography.

In 1964 I was approached by Dr. Popenoe concerning the need for supplementary work in cultural anthropology, particularly with regard to land-use practices of Kekchi migrants who were flooding into the region from Alta Verapaz. United States Office of Education research grant No. FH 5-148 enabled me to spend three months in Guatemala in the summer of 1965, and two years later additional support from the University of Florida's Center for Latin American Studies permitted me to revisit the country and to set aside the needed time for completion of the manuscript.

At Lake Izabal, a fishing cottage belonging to Mr. Chris Hempstead served as ongoing headquarters for researchers from the university. Most of their work was being conducted on a nearby Hempstead estate, Murciélagos, and it was initially suggested that I work there as well. Since my basic interest was in the land-use patterns of independent Kekchi migrant colonies, however, the small population and the controlled estate conditions of Murciélagos discouraged me. Instead of working there, I turned my attention to the largest and fastest growing spontaneous rural settlement in the *municipio*—Chichipate.

The pinpointing of such a settlement was made possible by access to original 1964 census sheets, ceded to me through the kind services of the National Director of the Census in Guatemala City. In addition to helping in the selection of the community, these census sheets provided general background material for it. This was later supplemented by data from the National Observatory and the Registro de Propiedad.

Upon my arrival in Chichipate, I was allowed to share the hut occupied by the local school teacher. From time to time I would return to research headquarters on Lake Izabal and there would exchange notes and experiences with Samuel Snedaker, a Ph.D. candidate in the Department of Botany, who was involved in an ecological project at Murciélagos. On Sam's suggestion, I began to collect botanical specimens, soil samples, and insect pests.

In working through my materials, I came to feel the need for aid in translation and in the reduction of Kekchi terms and texts to some sort of standardized form. On my return to Guatemala in 1967, the Seminario de Integración Social in Guatemala City helped me locate a pair of linguists who had been working with Kekchi for many years. These two persons, Ruth Carlson and Francis Eachus, were more than generous with their time and talents. Following their usage, x has been employed for the š sound, and all accents have been omitted since Kekchi terms are universally accented on the last syllable.

The process of final analysis involved many people. Mr. Snedaker did an analysis of soil texture, checked many botanical identifications, and prepared the land-use projection which is found in Section III. Daniel B. Ward of the University of Florida herbarium did an initial identification of plant materials; Howard V. Weems of the Florida State Museum, with the aid of Nell B. Causey of Louisiana State University, identified insect specimens; and the soils laboratory of the College of Agriculture, under the direction of James Nesmith, analyzed soil samples for available nutrients. Velva R. Rudd and José Cuatrecasas of the Smithsonian Institution identified certain plant materials not identifiable at the university herbarium. Jacques Barrau of Yale University identified other plants, and Paul C. Mangelsdorf of Harvard University gave identifications for maize samples. Barbara Denton of the Department of Anthropology at Yale generously shared insights gained from a prolonged archaeological reconnaissance of the Izabal region.

One is always primarily indebted to persons who first stimulate interest in his field of inquiry. In the present case, Harold Conklin, during my student days at Columbia University, introduced me to some of the basic concepts and methods of research in systems of cultivation. To a considerable extent, I have used his pioneer study on Hanunóo agriculture as a conscious model. The use of such a model helps overcome one of the major problems in comparative anthropology, i.e., the fact that seldom do two different investigators look at the same things in separate cultures. It does not eliminate the problem entirely, however. My report is much less comprehensive than was Conklin's, due in part to the fact that I spent less time in Chichipate than he did with the Hanunóo, in part to the fact that my background has been in the humanities rather than in the natural sciences, and, in part, I am convinced, to the fact that slash and burn agriculture in Chichipate is a decidedly less complex system than is slash and burn agriculture among the Hanunóo.

Ideally I should have spent a minimum of twelve months in Chichipate in order to observe a complete yearly agricultural cycle. However, responsibilities at the university made this impossible, and the best I could do to remedy the situation was to employ a variety of data-gathering techniques in an attempt to do some testing for accuracy. These techniques involved informational sources of varying types, and they should be clarified in order to insure the reader's accurate evaluation.

Original data sheets from the 1964 Guatemalan census provided basic information regarding the number of households in the community, the municipality from which individuals had migrated, the period of time each individual had lived in the *municipio,* his age, sex, and birthplace, and his relationships to other people in his household. This material was updated in 1965 when, by actual count of dwellings, the number of households was expanded and the specific locale, or *finca,* from which each person had migrated was ascertained.

My core source of data, however, consisted of ten household heads with whom I worked during the entire summer period. Information supplied by them includes data on agricultural stages, type of land used for milpa sites, typology of soils, clearing techniques, burning procedures, planting techniques, labor group membership, indigenous measures, weeding techniques, insect and animal pests, harvest procedures, granary building, storage, intercropping, dry season extension milpa cultivation, and the succession of specific plant types in secondary growth. Crude sketch maps prepared by these informants formed the basis for my updating of the 1964 census data. Similar sketches they made of surrounding milpas helped me to estimate average milpa size, as well as to arrive at percentage estimates of the various types of sites in which 1965 wet season milpas were located.

For purposes of obtaining information on patterns of land access, and the relationship between labor input, production, and consumption, I expanded my informant base to thirty household heads during the last two weeks spent in the field in 1965. This expansion allowed me to simultaneously cross-check many types of information I had already obtained from the core sources of data.

In my work with informants, I took great care to follow the open-ended approach, and basic cross-checking occurred in this fashion. Additional checks came from participant observation and from the fact that when the first draft of the manuscript was completed, I again traveled to Guatemala to go over spelling and translation of linguistic items, take more soil samples, and review the manuscript word-for-word with key informants.

When I first undertook this study, I had hoped that the reports of other investigators who have recently worked in the Izabal region would precede mine. Unfortunately, this has not proven to be so. Rather than being the last of a series of reports, mine turns out to be one of the first. As other material emerges, it may alter a number of my assumptions and conclusions. In the end, however, our knowledge of the agricultural potential around Lake Izabal should be as complete as that which exists for any humid, lowland, tropical region in the world.

William E. Carter

University of Florida, Gainesville

Contents

Section I. Pioneers in the Tropics

FROM ALL THE HIGHLANDS of tropical America, small groups of people are today migrating to the lowlands. This migration, stimulated by population pressure and encouraged by modern transportation, has steadily grown during the last decade. Migrants, on arriving at their destination, find that they must change a number of traditional ways. Climate, soils, vegetation, and even supernaturals differ from what they knew in the highlands. Crops and cropping techniques, dress, diet, religion, and social structure must all be adjusted accordingly.

This is an account of the adjustment one group of migrants has made, particularly with regard to crops and cropping techniques. More by accretion than by preconceived design, during the past six years the members of this group have carved a small but viable hamlet out of the Guatemalan rain forest. They chose their site, Chichipate, because it appeared to be in the public domain, it had an adequate water supply, it had both elevated lands and flatlands, and it was serviced by a main road which had just been opened between the lowland town of El Estor and the highland provincial capital of Cobán. The first pole and thatch huts were built at the point where a sizeable underground stream breaks out of the base of a hill and crosses the road. As the hamlet grew, it stretched along that road until today it has the form of a line town. The population has grown from zero in the 1950 census to 346[1] in the 1964 census, and this growth continues, making it one of the two largest, and certainly the fastest growing, of all rural settlements in the *municipio* of El Estor.

Migrants to Chichipate all speak Kekchi. Most of them come from Alta Verapaz, one of the largest but least densely populated departments of the Guatemalan highlands.[2] In moving to the lowlands they have reinstituted a migratory process that appears to have long historic precedent for the Maya region. However, the utopian hopes which most of them bring are quickly shattered by reality. They find the climate more demanding than they anticipated; temperatures climb to 100°F. or more any month of the year, and humidity is consistently high. The accompanying chart should supply sufficient background for understanding certain problems faced by the migrants, as well as the general timing of the agricultural cycle. Malaria-bearing mosquitos are ubiquitous, and in 1965 a malaria control unit found all but one member of the hamlet suffering from the disease. Drainage in large parts of the hamlet site is poor. Some house plots become quagmires in the rainy season. A few houses have been relocated to higher ground, but the residents of others have simply learned to tolerate muddy house floors.

Strong internal cohesiveness has never developed. A sizeable portion (37 per cent) of the migrants come from a single highland estate, but these people are so internally divided that they seldom present a united front. Inter-familial cooperation, where it exists, tends to follow previous lines of kinship or personal friendship, and thus to be limited in scope.[3] The teacher at the government school

1. All 346 were migrants or the offspring of migrants. No information is available on the number of persons who, after settling, have deserted the hamlet, though the general flow is into rather than out of it. By 1965 continued migration had swollen the population to 432, and by 1967 to 604.

2. Alta Verapaz shows both a lower population density and a lower rate of growth than does Guatemala as a whole. The country's population density in 1964 was 39 persons per square kilometer; that of Alta Verapaz was only 30. The country's rate of population growth between 1950 and 1964 was 3.1 per cent per annum; that for Alta Verapaz was only 2.4 per cent (Departamento de Censos, 1966: 14, 19). A factor accounting for this differential may possibly be a relatively higher rate of emigration from Alta Verapaz than from other highland departments. Kekchi peasants regularly leave Alta Verapaz to settle in Izabal and the Petén. For a detailed picture of the numbers involved, cf. Adams, 1965.

3. Forty-six per cent of the 346 migrants censused in 1964 had con-

finds such apathy on the part of parents that he must threaten to resign or denounce them to authorities in order to obtain needed repairs on the school plant. The two annually appointed auxiliary mayors encounter constant resistance to their orders.[4] Distrust and hostility are perpetuated through perennial malicious gossip, and social cleavages become particularly evident during the celebration of saints' days. Hamlet officials stand at the entrance to the fiesta grounds and collect machetes. They know that the men will try to outdo one another in drinking and inviting to drink, and they fear that if they do not collect the knives, several homicides will result from the inevitable bitter quarreling that ensues.[5]

The greatest sense of disappointment has come, though, from the fact that private absentee owners suddenly appeared after Chichipateños had cleared and settled the land. What the migrants had not known was that their site was part of a vast tract that was in litigation because of the bankruptcy of the Tinajas agricultural cooperative. They had mistakenly believed the land to be government-owned and thus open to squatters. Because of the litigation, during the first few years of settlement, no claimant appeared on the scene. Unbeknownst to Chichipateños, the litigation was finally resolved, and the land put up for sale. All that lay to the north of the El Estor-Cobán road was purchased by the International Nickel Company, while that to the south was purchased by the absentee owner

---

sanguineal relatives in the hamlet, and 62 per cent had affinal relatives. The preponderance of the latter type may reflect the custom of post-marital bride service, whereby a male is expected to live with and work for his wife's father for a variable period after taking her as a bride.

4. It may be that a lack of cohesiveness is characteristic of many agricultural settlements in the lowland Maya area. Reina, in writing of the Petén town of San José, suggests that the phenomenon is related to the nature of lowland subsistence activities: "The high degree of corporateness found in the highland Maya pueblos of Guatemala does not exist in San José. During the day the population of the town is composed mainly of women and children. During most of the year only 31 men (of 79 full time milperos) return daily from the fields to be with their families at night, and they depart before the sun rises. . . . It is evident that an elaborate cofradía organization [religious brotherhood: cf. Reina, 1966], if enforced in San José, would conflict with primary subsistence activities" (Reina, 1967: 10, 11).

5. It is striking how similar modern drinking patterns in Chichipate are to those described for the prehistoric Maya by Thompson: "After the meal the cupbearers served honeywine to the guests until they were helpless, when their wives appeared on the scene and conducted them home. . . . The Mayas when drunk were and still are a quarrelsome lot" (Thompson, 1927: 73).

of a neighboring estate, Yuscarán, on the banks of the Polochic River.

The demands made by the new absentee owners have been considerably lighter than they might have been. One common arrangement on estates in the Izabal region is to require the squatter to give

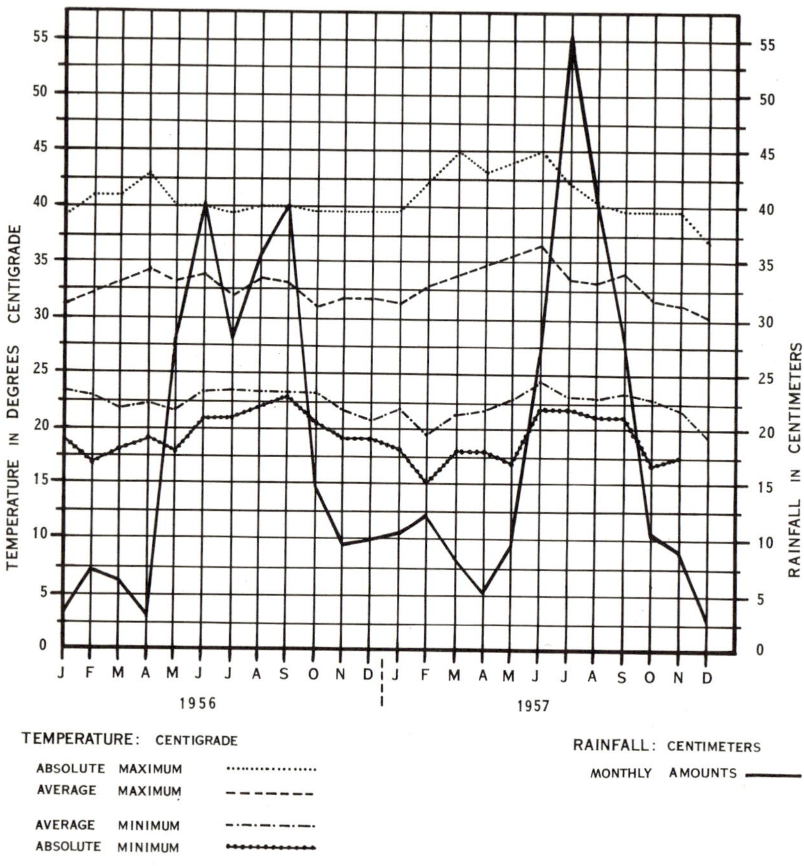

Temperature and rainfall measurements based on data collected in Yuscarán, an estate immediately adjacent to Chichipate.

two weeks of free labor per month during the periods of planting and harvest. Another is to require him to clear and cultivate approximately 60 per cent as much for the landowner as he clears and cultivates for himself. In the case of Chichipate, the man who took over the land to the south of the road was interested mainly in securing labor for his neighboring estate of Yuscarán, and clearing land so that he might eventually begin cattle raising around Chichipate itself. He therefore entered into an agreement whereby Chichipateños living on his property would work for him at their discretion, in return for which they could make milpa on the Chichipate land and would receive cash payment of fifty cents[6] per man-day of labor.

To the north of the road, the International Nickel Company made its purchase as part of a large-scale plan to begin extensive mining operations in this part of Guatemala. According to its locally stationed technicians and engineers, the company was disinterested in agriculture, except inasmuch as it threatened to lower the value of the purchase or to inhibit projected mining activities. It therefore turned over the administration of the squatters to the municipal government of El Estor in exchange for payment of a minimal fee by the municipality, a guarantee that slash and burn activities would be held within certain bounds, and an assurance that no more families would be permitted to settle in the hamlet. The municipal government appears to have seen this as an important new potential for revenue, and began to request a one-quetzal monthly rental fee from each family living on the company's land. When the fee was not paid, they began to conscript labor for some of the estates in the *municipio* that sorely needed it, but could not recruit it on a voluntary basis. The wages from such labor, instead of going to the conscripted laborer, were to go directly into the coffers of the municipality, until all back rental fees were paid.

Such an arrangement greatly displeased the Chichipateños. Many of them had migrated from the highlands for the very purpose of escaping from such obligatory services. So violent was their reaction to the situation that they temporarily overcame internal differences of opinion, solicited the aid of the school teacher, and wrote an appeal directly to the President of the Republic. The end result was that they were able to freely negotiate a yearly contract with the municipal government and, in 1965, were paying only fifty cents a month for fairly free use of land.[7] There was little indication that default of payment was bringing about any serious sanctions.

Certainly in the case of Chichipate, and probably in the case of

6. In 1965 the official rate of exchange between the Guatemalan quetzal and the United States dollar was one to one.

7. Substantially the same arrangements and course of events were described to the investigator by municipal authorities, mining company officials, and peasants.

all Guatemalan pioneer lowland settlements today, the rural community cannot be understood without taking into account its relationship with urban centers. The El Estor-Cobán road, finished in the early 1950's, now offers daily motorized transport to these two major centers of distribution for the Republic. Chichipateños living along the road collect and produce both agricultural and forest

The Cobán-El Estor road along which Chichipate has developed.

goods, in part, for the urban market. For both types of goods, prices are fixed by trends and events outside of the rural community.[8]

The fact that there is constant interplay between Chichipate and towns like El Estor and Panzós does not mean that Chichipateños will soon cease to be bearers of their own "little tradition." In spite of long and close contact with urbanized society, folk cultures tend to retain their distinct character and to resist change (Borhegyi, 1956: 352). George Foster has called them "half societies," parts of larger social units (usually nations) which are both vertically and horizontally structured (Foster, 1953: 163). Borhegyi has emphasized their symbiotic nature: "The folk component . . . bears a symbiotic spatial-temporal relationship to the more complex

8. Reina points out that even chicle must be sold at a price set by the agencies of the national government (Reina, 1964: 275).

component, which is formed by the upper classes of the preindustrial urban center. In this sense folk and urban are not polar concepts; rather they are both integral parts of the definition of a certain type of socio-cultural unit in which the preindustrial city is a focal point. Far from threatening the folk society, this type of urban unit is a precondition for its existence" (Borhegyi, 1956: 352).

In this light a comment revealing the world view of a Chichipateño in 1965 becomes particularly significant. During that year there had been violent guerrilla activity throughout the Department of Izabal. Rumors of massacres were frequent, and there was an uneasy quiet in the cities. When asked what he thought about all these events, the Chichipateño simply replied: "It doesn't really make any difference to me who is president, or whether the communists or the capitalists are in power. I live by my machete. As long as I have it, life, and some free land, I will get by." Yet this man long had been involved in national level institutions. Twenty years before he had migrated from Alta Verapaz to the Petén. There he had worked as a chicle gatherer and sold his products to government supervised agents. Later he had been employed by a mining engineer, and had journeyed throughout the highlands. Finally he began to gather chicle in the Chichipate area and, when the road was finished and squatters began to clear the forest, he settled down to milpa farming. A sizeable portion of his crop goes off every year to urban markets.

The open question, of course, is whether the lowlands will develop basically according to the traditional Guatemalan pattern represented by the case of Chichipate, or along the lines of modern, industrialized agriculture with peasants[9] being forced, in effect, into the position of a rural and urban proletariat. As the situation stands now, the peasants hold control over labor, and the elite hold control over land. Chichipate is a living example of the conflicts and stalemates which can ensue from such a situation.

The lowlands hold considerable potential for agriculture. Certainly in the past they were far more densely populated than they are at present, and debate is still going on as to whether aboriginal

9. The term peasant is used in the sense of cultivators who produce sufficient goods to meet practically all of their own subsistence needs, plus a small surplus which is siphoned off into regional and national markets. They are dominated by national institutions such as the government, the market economy, and the Church, though they exercise practically no control over any of these.

Maya culture originated there or in the highlands.[10] We do know that during pre-Classic times, small, more or less autonomous villages had developed throughout the lowland area. As the land was filled, ties between related villages appear to have become stronger, and ceremonial relationships were elaborated. From this came the centralization of Classic times (Dumond, 1961: 312).

The demographic profile accompanying this development is perhaps unique in the history of civilizations. At first only slightly settled by slash and burn agriculturalists, the lowland Maya region witnessed a spectacular population explosion during the Classic Period that resulted in a denser population than that of any part of Mesoamerica except certain highland regions where intensive agriculture was practiced. Within a century of the population peak, however, there was a sudden decline, and this decline persisted until the sixteenth century. Since then malaria and other European-introduced diseases have affected the region, and it has been nearly without population since Conquest (Sanders, 1962: 110-11).

The system of slash and burn agriculture is a strongly centrifugal one. Ordinarily it is argued that the inherent mobility of its component units will not lead to a strong state, based on tribute and labor drafts (Wolf, 1959: 60). Yet it was just such a system that seems to have produced the elaborate priestly centers of Classic times.

Many attempts have been made to explain the process of Classic development. Most authorities agree that the principal settlement pattern of Classic times was one with few or no congested centers of population, but with a scattering of dwellings almost anywhere that the land was not too swampy for habitation (G. Cowgill, 1964: 147). Some have pointed out that repressive authority seems not to have been a characteristic of the Classic Maya (Brainerd, 1954: 41, 70-75). On this basis, others have argued that Maya temples must have been the results of long series of accretions and that they are thus unlike the temples in highland Mexican centers such as Teotihuacán (Erasmus, 1965: 298).

A number of theories have been proposed to explain why the elaborate system of temple centers collapsed. One of the most extreme has attributed the collapse to severe earthquakes (MacKie, 1961: 216), though few have taken this seriously. More common

10. Reina has recently written: (there is an) "accumulation of archaeological evidence suggesting that the Maya culture originated in its tropical forest environment" (Reina, 1967: 1).

have been explanations related to the encroachment of grass on lands of cultivation. Recently, however, Ursula Cowgill's stratigraphic work in the Petén convinced her that grasslands there are clearly natural and not man-made. She found that soils in the savannas are quite different from those in the regions covered by forest and that, while forest areas are generally hilly and well-drained, savannas are flat and poorly drained. She also observed that, while there is probably no district in all the Petén which has been more continuously under cultivation than the areas near Lake Petén, and while these areas are being most heavily used at present, there is no tendency for grass to be a problem on them (U. Cowgill, 1962: 278).

As an alternative hypothesis to explain the end of the Classic, Ursula Cowgill suggests that a growth in population during that period could have led to temporary overstrain of agricultural resources which might then have been a factor in initial population decline, if not in its persistence (U. Cowgill, 1962: 279). Another explanation she offers is that population may have declined in part because of the evolution of a tradition tending to rejection, and hence less careful nurture and higher death rate, of female children. Yet another is that expansion of population and farmed areas may have meant a relative shortage of wild food rich in protein, possibly having a significant effect on Maya nutrition, physiology, and demography (Cowgill and Hutchinson, 1963c: 90-103). She concludes that the collapse of the Classic "may ultimately be solved in terms of a multiplicity of small and often non-obvious causes rather than a single dramatic catastrophe" (Cowgill and Hutchinson, 1963c: 101).

Even during the period of intensive Classic development, the territory immediately adjacent to Lake Izabal was never heavily settled. Ceramic remains indicate that the southeastern end of the lake served as a direct channel of communication between Copán and Quiriguá to the south, and Petén sites such as Tikal to the north. But recent archaeological reconnaissance has failed to uncover any indications of intensive settlement around that or any other part of the lake (personal communication from Barbara Denton, 1966).

The lack of settlement in this fertile and well-watered region remains a mystery. Obviously, with the decline of the Classic Maya centers, travel around the lake region either disappeared entirely or was seriously curtailed. The territory became one of the least populated of any in the New World.

During the colonial period, Lake Izabal came into play as the major Guatemalan port of entry for merchandise and passengers arriving from and departing to Spain. A fort was built at the point where the Río Dulce drains the lake, and at one time a heavy chain was strung across this stretch of water in a fruitless effort to protect Guatemalan commerce from pirateers (Brigham, 1887: 34-35). Population clusters began to form on the coast, but hardly penetrated inland. Highland peoples of the period showed no inclination to migrate into the Izabal basin.

With the development of Puerto Barrios, some miles to the south of the Río Dulce, and the building of a railway from Barrios to Guatemala City, the importance of the lake again decreased. United Fruit Company operations in this century have resulted in considerable agricultural development to the south of the lake, but to the north and west the territory continues to be one of the least developed and most lightly settled of any in Central America. According to the 1964 census, the Department of Izabal had only 13 inhabitants per square kilometer, while the country as a whole had 39. The municipality in which Chichipate is located, El Estor, had only 1.9 inhabitants per square kilometer. In Guatemala, only the Petén, with 0.8 inhabitants per square kilometer, had a lower population density (Departamento de Censos, 1966: 14).

Both the Petén and Izabal, however, are currently growing at a more rapid rate than the nation as a whole. From 1950 to 1964, the average annual population increase for Izabal was 5.1 per cent, and for the Petén it was 4.0 per cent, whereas for Guatemala as a whole, it was only 3.1 per cent (Departamento de Censos, 1966: 19). This indicates that the northern lowlands will soon be of more importance than they have been at any time since the Classic Maya period. George Cowgill has calculated that at the present rate of growth the Petén should reach a population of a quarter of a million in another seventy years (G. Cowgill, 1963: 500).

It may be said, then, that northern Guatemala today stands as a prime example of the resettlement of aboriginally occupied areas (Sauer, 1958: 106) of comparatively high subsistence potential (Ferdon, 1959). With all their potential, when unpopulated these areas have not been very strategic either economically or politically. Land shortage elsewhere may be what led to their occupation by cultivators in the first place (G. Cowgill, 1964: 154-55). It certainly is what is driving the Kekchi into the areas today.

It was once the fashion to accuse shifting cultivators of destroying vegetation and soils, and to blame poor returns on the method employed. Now it is fashionable to blame poor returns on the soils and to describe shifting cultivation as a wise adjustment to environmental conditions (Morgan, 1957: 9). Critics who have dis-

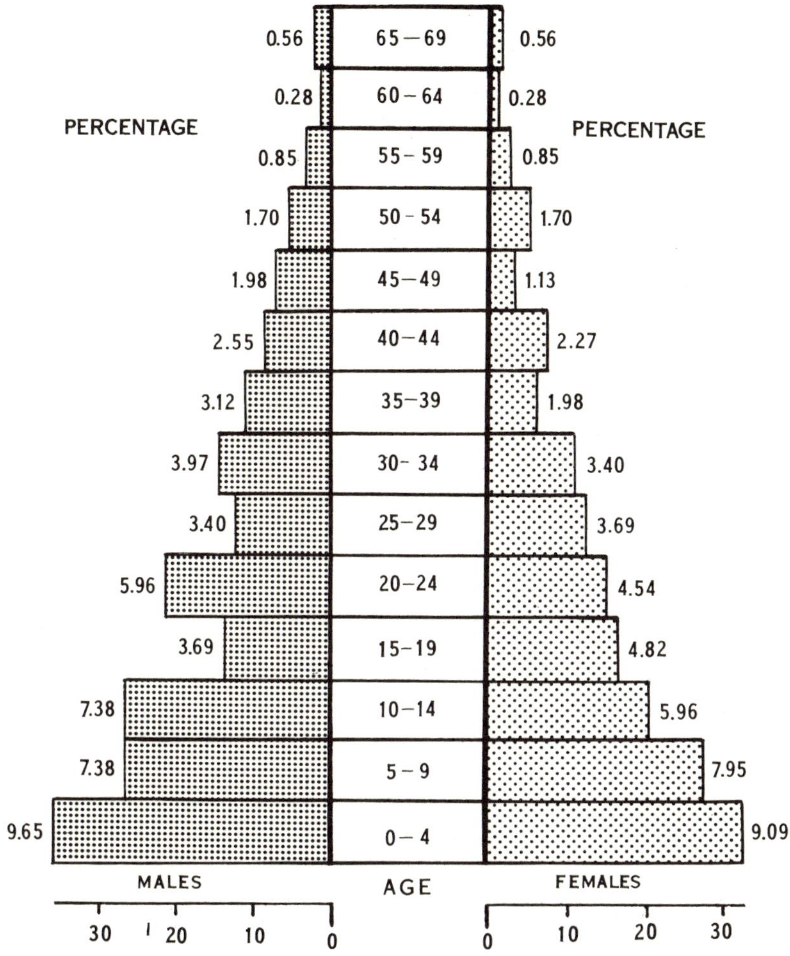

Age-sex pyramid, Chichipate.

missed shifting cultivation as a wasteful method have based their judgment on the destruction of timber and the dangers of deforestation and erosion (Leach, 1959: 64). That it does destroy forest cover is, to a certain extent, indisputable. In its better examples, however, it can be a conservative tool (Dumond, 1961: 305).

Popenoe has suggested that parts of the Polochic Valley, in which Chichipate is located, serve as just such an example; after considerable use, the soil continues to exhibit comparatively high levels of fertility (Popenoe, 1960: 76).[11]

Chichipate's age-sex pyramid indicates the futuristic orientation of modern migrants to the region. In 1964, 56 per cent of the population was under 20 years of age, 23 per cent was from 20 to 35 years old, and only 21 per cent was over 35. The fact that this youthful population has elected to remain in spite of difficulties of climate, disease, land tenure, and social hostility indicates that the lowlands do represent hope for highlanders living at the margin of their society. A detailed study of the adjustment these people have made should go far in helping to assess the developmental potentialities of the region. Much as the Guatemalan government should like to do so, it simply lacks the capital to promote large-scale, industrialized agriculture in the lowlands. The opening of new lands in the future is likely to follow rather closely a pattern similiar to that set down in Chichipate.

11. The reader is referred to Popenoe's Ph.D. dissertation (Popenoe, 1960) for both an excellent introduction to the geology and geography of the Polochic Valley and a thorough discussion of differing views on the functions and merits of shifting cultivation.

Section II.
The Milpa Cycle

Since the hamlet of Chichipate is a new one, agriculture is still in many ways in the experimental stage. A few of Chichipate's residents have lived in the lowlands as many as twenty years, though most came from higher regions during only the past six years. The earliest arrived before the El Estor-Cobán road was built, and were engaged in the collection of chicle. Others came when the road was being put through, and were, in effect, road workers. Because of their years of experience, these older residents are sought by the newcomers as experts on regional agriculture. It is they who carry on the greatest plant experimentation and grow the widest variety of cultigens.

All residents, new or old, produce agricultural goods for both subsistence and the national market. Maize is the crop par excellence, and sizeable amounts of it are sold at the end of each growing season. With proceeds from these sales, Chichipateños purchase foodstuffs, such as beans, to which they are accustomed but which they can grow only with great difficulty in their new environment. Middlemen who purchase their produce and who bring in outside items tend to be Ladinos from municipal or departmental capitals. Some live permanently in El Estor, the nearest of such capitals. Others come from as far away as Cobán, or even Guatemala City.

The dry season crop, harvestable in March and April, is in particularly high demand. Its harvest comes at a time when supplies of maize in highland Guatemala are low. Whereas wet season maize will sell for from Q1.50 to Q2.00 a *quintal,* dry season maize occasionally brings Q6.00.

Production for sale and trade is decidedly secondary to the milperos' interest in obtaining food for themselves and their families. Chichipateños are anything but omnivorous. They grow and eat rice and different types of squash, and they regularly import a number of fruits. But these items are marginal. Indeed, some families never plant bananas, even though they live in an excellent banana-producing area. The real staples of the diet are corn and beans.[1]

During my stay in Chichipate, I made a habit of asking people what they had for breakfast or lunch or dinner. In almost every case they would reply: "I had chili" or "I ate squash leaves" or "I had manioc." Hardly ever did they mention the fact that, along with their squash, or their manioc, or their greens, they ate quantities of maize tortillas, tamales, and gruel-like *atole.* These maize dishes are, apparently, so omnipresent in their diet that they do not ordinarily think about them. They are not the delights of life; they are the essence of life.[2]

1. The nutritive qualities of maize are impressive. Hester has written: "It is high in carbohydrates, but relatively low in proteins and fats. Souza-Novelo (1950: 165-166) provides the following general analysis of dry Yucatán maize: moisture, 1.40 percent; ash, 1.41 percent; protein, 9.90 percent; fats, 4.4 percent; crude fiber, 1.97 percent; total assimilable carbohydrates, 71.92 percent. Deficient or lacking in some of the more important vitamins and minerals, maize alone would not suffice to support an individual through a normal human life cycle. But supplemented by animal and other plant products, the desirable qualities which maize offered to the Maya diet far offset its dietary deficits" (Hester, 1954: 152).

2. Nutritional studies are still lacking for this part of Guatemala. Nutritional experts working in Yucatán, however, have found that maize prepared in various ways accounts for an average of 73 per cent of the daily energy intake, and in some instances supplies as much as 86 per cent of the total (Benedict and Steggerda, 1937: 184, 188). Hester has hypothesized that it is probably safe to assume that maize products provide a minimum of 50 per cent of all food consumed in the lowlands (Hester, 1954: 72-73). After his years of work with the Yucatecan Maya, Morley was so impressed by the potentialities of maize that he described it eloquently as "nature's richest gift to man—the Maya staff of life—without which they never could have developed their highly distinctive culture, the most brilliant aboriginal civilization in the New World. . . . Their agriculture was based directly upon, and derives straight from, agriculture as applied to the cultivation of corn. [This is] . . . the most basic fact about Maya civilization" (Morley, 1947: 158).

In homes of the Chichipate Kekchi, families seldom sit and eat together. Each individual eats as he feels the urge, and this may be at any hour of the day or evening. Men often sit alone in a hammock drinking fish soup, or eating corn on the cob, or scooping up beans with maize tortillas. Occasionally people are seen eating in the doorway of their house, especially if this is shaded from the hot tropical sun. The woman of the family may announce that food is

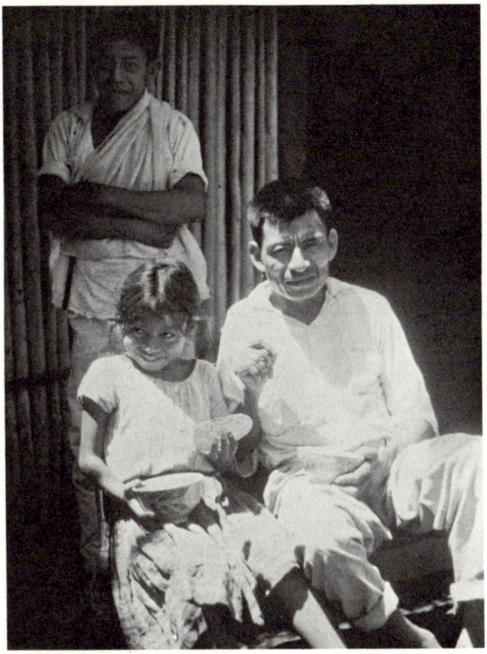

Occasionally people are seen eating in the doorway.

ready. There is, however, never any rush to the table. One eats according to his own desires and needs.

Relatively few utensils are used in eating. Most families have spoons, but much food is simply picked up with the fingers. Women do not exert great efforts in preparing elaborate dishes, and much Kekchi food would be considered by us as tasteless and without seasoning.

The primacy of the maize staple contrasts sharply with the concept of a meal among less vegetarian people like ourselves. Maize is the most important real food. It is indispensable for rites, offerings, and for both large feasts and small meals. Individuals may experiment with other plants, but it is their success with maize that determines their standing in the community. Many milpas are planted to this and only this crop. The entire agricultural cycle is geared to maize cultivation. Yet only five varieties of this plant are cultivated in the community,[3] and of these only two are anything like common; the others are cultivated by only scattered individuals. Some varieties, black for example, are grown mainly for ceremonial purposes.

In comparing foods grown in Chichipate with those grown in a region like the upland Philippines, one cannot but be impressed by the lack of variety of the former. This may be due in part to the fact that Chichipate is a pioneer hamlet, though there is the distinct possibility that slash and burn agriculture, as practiced by the Maya, was never as complex as it is in other parts of the world. One thing that makes it difficult to judge whether this is so is that we know the lowland Maya have, since Conquest, lost specific cultural attributes, such as, for example, sailing.

The system of slash and burn, or shifting cultivation, is commonly known in Mesoamerica as milpa agriculture. The term seems to come from Nahuatl, and simply means "cornfield" (Morley, 1947: 141). In Kekchi, the word for cornfield is *c*ˀ*al*, and similar terms exist in all the Maya languages. Chichipateños, however, recognize and use the word "milpa" when conversing with Spanish speakers.

This system of cultivation, variously called milpa agriculture, swidden farming, digging stick or hoe agriculture, slash and burn, and shifting cultivation, is worldwide in distribution, being found especially in forested tropical lowlands (Lundell, 1937: 11).[4] No matter how limited the variety of crops grown under the system, when distance, acreage, nature of help, type of soil, slope, weather conditions, and beliefs are considered together with age, health, or personal preferences, the milpa system emerges as a highly complex economic and sociological phenomenon (Reina, 1967: 14). For analytical purposes, this phenomenon can always be broken down,

3. In highland Guatemala, many more varieties of maize are commonly grown than in the lowlands. The difference may be due to the fact that the Izabal region presents a much more extreme environment for maize, to which very few varieties are well adapted.

4. For good descriptions of the Middle American system, see Cook (1921), Lundell (1937), Steggerda (1941), and Kelly and Palerm (1950). For an exhaustive bibliography of published references of shifting cultivation in general, see Conklin (1961).

however, into a limited number of phases and processes. In his study of Hanunóo agriculture, Conklin has argued that the phases are basically two: removal of old vegetation and control of new vegetation. Within these two basic phases, he sees five general processes (Conklin, 1957: 31): site selection, cutting, burning, cropping, and fallowing. Conklin's two basic phases, being mere analytical categories, are relevant to though not overtly recognized by the Kekchi of Chichipate. In contrast, the five general processes are discussed in daily conversation, though not designated with the terminological specificity he encountered with his Hanunóo. Within these five general processes, the Kekchi recognize a number of specific stages which they identify by small descriptive phrases (Table 1).

In the Chichipate system, greatest emphasis is placed on wet season cultivation in fields freshly cleared from land that had previously been either in fallow or high forest. Dry season plots tend to be smaller, less productive, and hence of minor importance. Wet season fields are often intercropped with isolated plantings of marginal foods such as manioc, malanga, bananas, sweet potatoes, sugarcane, and pineapple. But these are seldom weeded after the second or dry season extension maize crop has been harvested. Under these conditions, seldom do intercropped items last more than two to three years. The single exception is pineapple, small plots of which may hopefully be kept productive for as long as ten to fifteen years. Population pressure has not yet pushed people to realize the full potentialities of intercropped materials.

Within each normal wet season stage, Chichipateños recognize different substages of development related mostly to the cycle of maize maturation. These will be discussed when the actual procedures of cropping are considered.

Site Selection

Chichipate, in 1965, had ninety separate households whose members loaned and borrowed labor regularly among themselves. However, it was the household unit, the core of which was the nuclear family, that formed the basic work group. There appears to have been no insistence that each adult male have his own milpa, and a number of young married men simply made milpa with their fathers and/or brothers. There was concern, however, that each family produce enough for sale as well as for food. Older settlers of the community had already preempted much of the land. As a

TABLE 1

MILPA STAGES

Kekchi	English
Specific Normal Stages	
Al c'al	Milpa resting
C'alenbil c'al	Slashed milpa
Ak'inbil c'al	Cleaned milpa
Inc'a' c'atbil li c'al	Felled but unburned milpa
C'atbil li c'al	Burned milpa
Aubil li c'al	Seeded milpa
Xmok li c'al	Plants broken through the ground ("milpa is pointed")
Mas li pim sa' c'al	Weedy milpa
Ak'inc li c'al	Weeded milpa
Xk'ano' li c'al	Mature milpa, silk dry, stalks green
Xchakic li c'al	Dry milpa (ready for harvest)
Chamal pim	Low fallow milpa (thick)
Coc' che' ru	Sapling stage of secondary growth
Nīnk al c'al	High fallow milpa
Nīnk li q'uiche'	High forest
Subsidiary Stages*	
C'alenbil sak'icuaj	Slashed for dry season extension milpa (slashed for corn of the sunny season)
Sak'icuaj	Dry season extension milpa (corn of the sunny season)
Cuan li acuīmk sa' li rok cuaj	A few productive plants in the fallow (there are productive plants in the old corn stalks)
Disjunctive Stages	
X'oso' li cuaj	Ruined (young plants are finished, e.g., because of wind, weeds, or insects)
Pach'aya'	Grassland

*More intensive seasonal cropping has recently been reported for the Petén. In San José, milperos may sow three crops in one year, that is, a regular crop of first or second year milpa and two emergency crops: San José, a dry season crop seeded in low lying sites; and Yaxkin, seeded during the rainy season on higher terrain (Reina, 1967: 1).

result, a number of households cultivated two or three small plots, rather than one large one. For ninety households there were one hundred and thirty wet season milpas.

A number of variables enter into the selection of a site for conversion to milpa. These will be considered in the order that the Kekchi perceive them.

Freedom of Action

The biggest single attraction that the lowland tropics hold for the highland Kekchi is freedom of action. If they moved into the Chichipate area, it was because when they did so they believed it to be government land. The fact that it really belonged to the Tinajas cooperative, and that it was recently split and sold, one part to a mining company and the other to a Guatemalan entrepeneur, has proved to be a bitter disappointment. Some families have left, and others are threatening to do so. The only reason they stay is that the mining company, through the administration of the municipal government, has made the rental fee minimal, and the Guatemalan entrepeneur has proven to be benign. It may be safely assumed that had the owners of the Chichipate area been as active in 1961 as they are today, the hamlet would never have been established. As soon as undue pressure is exerted by either of the present owners, the place will undoubtedly be deserted, and the inhabitants will migrate elsewhere, likely squatting on what they presume to be other government lands.

Previous Use of Land

Migrants coming to Chichipate carry with them the belief that the best lands are those which have never previously been farmed.[5] Were they to follow this ideal, they would use a great deal of mature forest land for their wet season milpas. However, the mandates of the new landlords forced them to compromise and, in 1965, they were using principally plots that had been farmed previously. Sixty-eight per cent of their wet season milpas was in lands that had been allowed to lie fallow not more than three years, 21 per cent in lands that had lain idle for four to ten years, and only 11 per cent in lands that had just been cleared from mature forest growth—the last

5. This ideal of searching for virgin lands for cultivation is unusual. In Yucatán the best milpa lands are thought to be the opposite—former village sites and previously farmed areas (personal communication from Roland Chardon, 1968).

often in direct opposition to mandates from the landlords. There was considerable complaint on the part of the residents over the dictum that no more virgin timber be felled, and many saw this as the toll of doom for the community.

Though intercropping is practiced and results in small stands of manioc, sweet potatoes, chili peppers, malanga, pineapple, and sugarcane that can survive for several years after the initial milpa harvest, these crops are considered relatively unimportant. They seldom cause a person to hesitate before slashing or burning a field, particularly after the second year of fallow growth. With the exception of pineapple, no intercrop is thought to be worth even desultory weeding after the third year.

Because of a national law prohibiting the destruction of mahogany, Chichipateños insist that they never use lands where these trees stand.[6] Observation, however, indicates that such lands are actually used, the trees sometimes being destroyed and other times being left as lone specimens in the middle of the milpa. A second national law which prohibits the felling of trees of more than 6 inches in diameter for milpa is disregarded entirely.

Though tenure is by usufruct, once an individual has cut and burned over virgin land, he establishes a claim to it in perpetuity. More recent arrivals to Chichipate thus are forced to request permission from older residents before entering areas of secondary growth. Since these older residents settled at a time when there were no restrictions on land use, they have extensive fallow holdings and, in 1965, gave access quite readily. A substantial increase in population would rapidly make the continuance of this practice difficult, however.

Types of Soil

The fact that Maya lowland soils are not particularly poor has been suggested by the anthropologist Ferdon (1959: 13). He is supported by various soil studies which have been made in the region (e.g., Wright *et al.*, 1959; Cowgill and Hutchinson, 1963a; Popenoe, 1959). Simmons, Tárano, and Pinto's soil maps (1959)

6. The same regulations affect all land clearing in lowland Guatemala. In San José, Reina found that municipal authorities enforce the forestry regulations. If there are large trees in the field that can be used for building canoes or houses or for chicle gathering, a milpero must secure official permission to use them, he must be certain that he does not exceed his quota, and he must pay the set tax for each tree (Reina, 1967: 4).

indicate that, for the Chichipate area, Sebach soils predominate, though Setal and Semuc soils are also present in spots. Sebach soils are described as dark brown, friable clays approximately 30 cm in depth, deriving from and overlying serpentine. Setal soils are derived from alluvium, are considerably deeper (50-60 cm), and may be classed as reddish brown, friable clays. Perhaps because they occur on flatlands, they tend to present more drainage problems than do Setal soils. Semuc soils are also reddish brown, derive from and overlie serpentine, are 60-80 cm deep, and have an excellent clay base structure (Simmons, Tárano, and Pinto, 1959: 545-52).

Chichipateños are very aware of soil quality; they consider it primary in the production of good crops. Their criteria for classification are mainly four: color, texture, drainage, and root content. On this basis they devise eight major categories for color, ten for texture, five for drainage, and one for root content. In Tables 2 and 3, these are set down *in the Kekchi's own descending order of quality, going from the best and most productive to the worst and least productive*. The eight color categories are often combined with categories of texture and condition in describing specific soils. However, just as frequently, they are considered independently as diagnostic signs of soil quality.

The categories for texture are anything but mutually exclusive. *Samahiʔ, muʔ, pec,* and *pok chʔochʔ* are almost always soft-surfaced soils and so fall into the generalized textural category of *kʔun,* or soft. *Melb ru li chʔochʔ* may be considered either soft *(kʔun)* or hard *(cau),* depending on whether the hardened clay has been fragmented or not. *Seb, saclun,* and *cuācab ru li chʔochʔ* are usually typed as *cau,* especially when they lie on the surface of the ground.

In addition to falling into general textural categories, the eight specific soil textures tend to be associated with specific soil colors. Thus, *pec ru li chʔochʔ* may be shades of black *(kʔek),* yellow *(kʔan),* or red *(cak),* but it cannot be white *(sak). Samahiʔ ru li chʔochʔ* is found only in shades of black *(kʔek),* and especially in brownish-black *(mero kʔek). Muʔ ru li chʔochʔ* is found in all four basic colors: black *(kʔek),* white *(sak),* yellow *(kʔan),* and red *(cak).* Hardened clay *(melb ru)* comes in yellow *(kʔan),* light brown *(mero kʔan),* red *(cak),* and white *(sak),* but not in tones of black. *Saclun* may be found only in white *(sak)* and yellow *(kʔan)*; *seb* comes only in red *(cak)* and yellow *(kʔan). Cuācab* comes only in dark browns and brownish-blacks *(mero kʔek* and *kʔek),* and *pok* comes only in white *(sak).*

Kekchi classifications of texture, though they do not closely correspond to those established by modern soils scientists, seem perfectly adequate for milpa farming. The important thing the milpero is interested in, with regard to texture, is how coarse or fine the soil might be. He prefers the coarse, well-aerated type. This he can select quite skillfully by surface examination alone.

TABLE 2
SOIL COLORS

Kekchi type	English designation and usability
K'ek li ch'och'	Black soil. Very good. The most widely distributed of good soils.
Mero k'ek li ch'och'	Brownish-black soil (half black). Good for both wet and dry season milpa. Widely distributed.
Mero k'an li ch'och'	Brownish-yellow soil (half yellow). If soft *(k'un)*, good for both wet and dry season milpa. Limited distribution.
K'an li ch'och'	Yellow soil. If soft *(k'un)*, good for both wet and dry season milpa. Often, however, it is hard *(cau)*. Limited distribution.
Sak li ch'och'	White soil. If soft *(k'un)*, good for both wet and dry season milpa. If hard *(cau)*, it is very poor for milpa. Limited distribution.
Mero cak li ch'och'	Reddish-brown soil (half red). Fair for milpa, especially if soft *(k'un)*. Very widely distributed.
Cak li ch'och'	Red soil. Fair for milpa, especially if soft *(k'un)*. Very widely distributed.
Tzakal cak li ch'och'	Very red soil. Poor for milpa. Fairly widely distributed.

The textural categories most common in Chichipate are: *samahi' ru*, *melb ru*, *mu' ru*, *seb ru*, *pec ru*, and *cuăcab*. In an attempt to gain some insight into the semantic fields of these categories, six soil samples were taken for each of them, all being selected on the basis of surface texture alone, with the exception of a *seb* sample which was chosen because it bordered on a well-known source of clay. Informants who helped collect the samples classed those of *samahi' ru*, *melb ru*, and *mu' ru* as soft *(k'un)* and the rest as hard *(cau)*.

TABLE 3
SOIL TEXTURE

Kekchi type	English designation and usability
K'un ru li ch'och'	Soft surfaced soil. Good for both wet and dry season milpa. Widely distributed.
Samahi' ru li ch'och'	Sandy, silty loam. Excellent for wet season milpa if well drained. Good for dry season milpa. Limited distribution.
Mu' ru li ch'och'	Very soft surfaced soil. The top layer of this soil tends to be partially decayed organic material. It is the best soil for dry season milpa, especially if it lies on low, flat land. Less productive for wet season milpa. Limited distribution.
Melb ru li ch'och'	Surface covered by pieces of hardened clay. Good for both wet and dry season milpa. Especially productive for beans. Limited distribution.
Pec ru li ch'och'	Stony surfaced soil. Usable for wet season milpa if the stones are whitish or bluish and if covered with ligneous growth. Stony soils which produce only sparse herbaceous growth or grasses are eliminated from consideration. Common on hilly sites.
Saclun ch'och'	Yellow and grey clays, usable for dry season milpa if near water. Rare in Chichipate.
Seb ru li ch'och'	Yellow and red clay surface. Excessively hard in the dry season and sticky in the wet season. Poor for milpa if the surface of the ground is *seb*. *Seb* may lie under softer, surface soils, however, and in this case produce good milpa. Fairly common in Chichipate.
Cuăcab li ch'och'	Black and brownish-black clay. Similar to *seb,* and presenting many of the same problems. Rare in Chichipate.
Pok ch'och'	Partially decomposed limestone, used as a cleanser. Poor for milpa, though rare in Chichipate.
Cau ru li ch'och'	Hard surfaced soil. Poor for both wet and dry season milpa. Fairly widely distributed.

It will be remembered that this may not always be the case, particularly with *melb* and *pec* soils, which may be either hard or soft. However, the results of a laboratory analysis using the hydrometer method (Bouyoucos, 1936) clearly illustrate why the informants so classified these samples (Table 4). The *melb* soils contained sizeable

TABLE 4
KEKCHI TEXTURE CLASSIFICATION AND ANALYSIS

Kekchi type	Depth (inches)	Per cent of sand	Per cent of silt	Per cent of clay	English type
Samahiˀ ru	0-2	63.1	18.0	18.9	Sandy clay loam
li chˀochˀ	2-6	57.2	19.9	22.8	Sandy clay loam
	6-10	59.9	15.5	24.6	Sandy clay loam
Mean all levels		60.1	17.8	22.1	Sandy clay loam
Muˀ ru	0-2	51.6	24.5	23.9	Sandy clay loam
li chˀochˀ	2-6	52.7	21.5	25.7	Sandy clay loam
	6-10	35.9	25.4	38.7	Clay loam
Mean all levels		46.7	23.8	29.5	Sandy clay loam
Melb ru	0-2	48.9	27.2	23.9	Sandy clay loam
li chˀochˀ	2-6	32.1	32.3	35.6	Clay loam
	6-10	37.2	33.5	29.3	Clay loam
Mean all levels		39.4	31.0	29.6	Clay loam
Seb ru	0-2	30.1	34.8	35.1	Clay loam
li chˀochˀ	2-6	8.4	21.8	69.8	Clay
	6-10	0.0	15.0	85.0	Clay
Mean all levels		15.4	25.6	59.0	Clay
Pec ru	0-2	18.9	23.5	57.6	Clay
li chˀochˀ	2-6	19.4	16.0	64.6	Clay
	6-10	13.8	15.6	70.6	Clay
Mean all levels		21.0	18.2	60.8	Clay
Cuăcab	0-2	18.5	20.6	60.4	Clay
li chˀochˀ	2-6	11.6	18.2	70.4	Clay
	6-10	11.9	15.1	73.0	Clay
Mean all levels		14.2	17.4	68.4	Clay

portions of both sand and silt, while the *pec* soils, taken from a limestone-covered hill, turned out to have a very high clay content.

Overriding the importance of soil color and texture are the matters of drainage and root content (Table 5). A soil which may meet basic criteria for being of top productive capacity, i.e., black and soft, can nevertheless fail because of flooding or a root problem, and soils that are highly productive in the rainy season may dry or crack in the dry season and thus become worthless.

Since Kekchi designations for drainage are flexible depending on the moisture content of the soil at the time of observation, it is impossible to estimate the percentage of Chichipate soils that fall into

TABLE 5
DRAINAGE

Kekchi type	English designation and usability
Chaki ch'och'	Dry soil. Good for wet season milpa; poor for dry season milpa.
Sulul li ch'och'	Muddy soil. If the condition is only temporary, soil may be usable for either wet or dry season milpa. It is inferior, however, to soils having soft, granular surfaces. In general, *sulul* is more usable for dry than for wet season milpa.
Xjore' li ch'och'	Cracked soil. Cracks appear only during the dry season. Soils that crack are usually found on high ground or hillsides, and so are usable only for wet season milpa.
Ha' ru li ch'och'	Soil covered by water. Very poor for wet season milpa, though possibly good to excellent for dry season milpa. These soils are commonly derived from clays *(seb)*.
Sāb ru li ch'och'	Perpetual swamp. Useless for both dry and wet season milpa.
	ROOT CONTENT
Tzatzalum	Root-bound soil. Found almost exclusively in recently slashed and felled milpas, especially when they have been carved out of mature forest *(nīnk li q'uiche')*. These soils generally are poor the first year but, as the roots slowly rot away, become good to excellent by the second year of use.

each category. Furthermore, drainage and root content overlap color and texture in such a way that a lengthy series of soil types becomes possible: e.g., muddy white soil *(saki sulul li ch?och?)*; black, root-bound soil *(tzatzalum k?ek li ch?och?)*; and stony, yellow soil *(pec ru k?an li ch?och?)*. Judgments as to the merits of a specific soil depend on an astute evaluation of its various, sometimes contradictory, qualities, and social pressure encourages care in this evaluation. As Reina has indicated, the poor selection of a piece of land and the spending of one's energy for little or no profit are considered socially embarrassing; they are signs of poverty and stupidity (Reina, 1967: 3).

Independent from the already mentioned analysis for texture, a separate one was made to determine the relationship of specific Kekchi soil classifications to mineral nutrient content. This was based on 125 samples taken in 21 different milpas. Each of these milpas was sampled at two points, the control being that the soils at both points be given the same terminological designation by the Kekchi cultivator. At each of these points, subsamples were taken from three depths: 0 to 2 inches, 2 to 6 inches, and 6 to 10 inches. There resulted thirteen composite Kekchi soil types. In Table 6, presenting data from the laboratory report, these types have been *ranked in a descending order of quality, as judged by Kekchi standards.*[7]

The range in chemical content for the various samples representing any one soil type was small, and content figures have therefore been presented in averages. In general, the analysis indicates that Chichipate has fairly fertile soils. Other than questions of drainage, what problems do exist seem limited mainly to the fact that some soils are low in phosphorus, some low in nitrogen, and some have a high ratio of magnesium to calcium.

The rank ordering of soil types characteristic of the Chichipate system corresponds roughly to that Reina obtained from San José natives (Reina, 1967: 1-2). Chichipate informants place black

7. To arrive at the figures on chemical content, several techniques were employed. A Sargent pH meter, general purpose glass electrodes, and a 1:2 soil-water ratio were used to determine pH. Soils were extracted with 1N ammonium acetate buffered to pH 4.8. Calcium and potassium were determined on the extract by the use of the Beckman B spectrophotometer with flame attachment and photomultiplier. The phosphorous determination employed the use of a photovolt colorimeter, using the molybdenum blue procedure. Nitrate nitrogen was determined qualitatively by the diphenylamine method.

TABLE 6

CHEMICAL ANALYSIS OF SOIL NUTRIENTS

(Oxides in pounds per acre)

Soil name	pH	CaO	MgO	K_2O	P_2O_5	NO_3*	Harvest†	Rank in harvest
I								
1. *K?ek li ch?och?*	7.10	8641	7101	702	45.5	2.5		
2. *K?an li ch?och?*	6.25	1104	5621	433	0.75	2.00		
3. *K?an li ch?och?*	6.81	1045	7426	339	0.25	1.00		
X (black surface with brown underneath)‡	6.72	3597	6716	491	15.5	1.83	30	1
II								
1. *K?ek li ch?och?*	6.47	7578	2979	721	49.8	3.25		
2. *K?ek li ch?och? seb ru*	6.14	3367	1745	138	0.63	1.25		
3. *Seb ru*	6.17	2740	1883	121	trace	1.25		
X (black earth with clay underneath)	6.26	4562	2202	327	16.81	1.92	30	1
III								
1. *Samahi? li ch?och?*	7.56	4191	6312	344	3.8	2.00		
2. *Samahi? li ch?och?*	7.36	2448	6899	360	6.3	1.00		
3. *Samahi? li ch?och?*	7.13	1721	6453	259	0.75	1.00		
X (sandy loam)	7.35	2787	6555	321	3.6	1.33	30	1
IV								
1. *Melb ru*	6.92	5762	2640	560	25.5	1.5		
2. *Melb ru*	6.40	4550	3696	392	7.5	1.00		
3. *Melb ru*	6.04	3939	3850	284	1.75	1.00		
X (hardened, fragmented clay)	6.45	4751	3395	412	11.52	1.17	25	4

TABLE 6—*Continued*

Soil name	pH	CaO	MgO	K_2O	P_2O_5	NO_3*	Harvest†	Rank in harvest
V								
1. *K?an li ch?och?*	6.67	2113	1924	722	7.5	1.00		
2. *K?un li ch?och?*	5.75	834	1890	304	1.00	1.00		
3. *Chak li ch?och?*	5.45	500	2572	243	0.5	1.00		
X (soft, dry yellow earth)	5.96	1149	2129	423	3.00	1.00	25	4
VI								
1. *Mu? ru*	6.90	2560	5150	1047	7.5	1.00		
2. *Mu? ru*	6.37	1106	4982	560	0.5	1.00		
3. *Mu? ru*	6.61	640	6899	283	0.5	1.00		
X (trashy soil)	6.63	1435	5677	630	2.8	1.00	24	6
VII								
1. *K?ek li ch?och?*	6.99	5932	2212	1042	13.3	1.33		
2. *Cak li ch?och*	6.07	3081	1872	411	0.5	1.00		
3. *K?an li ch?och?*	5.89	2961	2324	337	.66	1.33		
X (black earth with red and brown underneath)	6.32	3991	2136	597	4.82	1.22	20	7
VIII								
1. *Sak li ch?och?*	7.05	4774	2011	972	37.5	1.5		
2. *Sak li ch?och?*	6.37	1616	2785	489	1.5	1.00		
3. *Sak li ch?och?*	5.94	1319	4361	354	.75	1.00		
X (grey or whitish earth)	6.45	2570	3019	605	13.2	1.17	20	7
IX								
1. *Cak li ch?och?*	6.50	3496	2025	668	2.00	2.00		
2. *K?an li ch?och?*	6.13	854	1681	364	trace	1.00		

3. *K?an li ch?och?*	5.40	476	1458	364	0.5	1.00		
X (red earth with yellowish-brown underneath)	6.01	1602	1721	465	.83	1.33	20	7
X								
1. *K?ek li ch?och? tzatzalum*	7.45	6848	5674	2038	142.8	2.5		
2. *K?ek li ch?och? tzatzalum*	6.52	1151	6453	560	1.5	2.00		
3. *K?ek li ch?och? tzatzalum*	5.63	986	8009	392	.5	1.5		
X (root-bound black soil)	6.53	2995	6712	997	48.3	2.00	17.5	10
XI								
1. *Mero cak li ch?och?*	6.59	7902	3571	932	22.5	1.5		
2. *Mero cak li ch?och?*	6.31	1249	1721	412	1.00	1.5		
3. *Mero cak li ch?och?*	5.94	441	1684	324	0.3	1.5		
X (half reddish soil)	6.28	3197	2325	556	7.9	1.5	15	11
XII								
1. *K?an li ch?och? ha? ru sulul*	7.08	5738	2491	1196	26.0	1.00		
2. *K?an li ch?och? sulul*	6.44	2427	2491	560	3.0	1.00		
3. *K?an li ch?och? sulul*	5.98	2126	4364	358	trace	1.00		
X (yellow earth; muddy, covered by water)	6.50	3430	3115	705	9.66	1.00	5	12
XIII								
1. *Seb ru ha? ru*	9.29	3797	44,766	1347	31.0	1.00		
2. *Seb ru*	7.90	3564	6497	2043	11.8	2.00		
3. *Seb ru*	7.56	3046	5279	1198	9.0	1.5		
X (clay, covered by water)	8.25	3468	18,847	1529	17.3	1.5	0	13

*1.00—very low; 2.00—low; 3.00—medium; 4.00—high.

†*Quintales* per *manzana.*

‡Words in parentheses indicate a rough English equivalent of the three Kekchi soil levels. Numbers to the right of this English equivalent represent averages for all three soil levels.

porous soils at the top and poorly drained soils, regardless of color, at the bottom. Next in preference after black soils are yellow, followed by white and red. Of the soils having no drainage or root problems, *mero cak li ch?och?* (very dusky red on the Munsell color chart) is ranked as the poorest.

The overriding importance of drainage is illustrated by the fact that, of the four lowest ranked soil types, all but one (No. XI—*mero cak li ch?och?*) have serious drainage problems. The composite

Maize in *sulul li ch?och?*.

Maize in *k?ek li ch?och?*.

types ranked twelfth and thirteenth are regularly covered by water *(ha? ru)*, and that ranked tenth is root-bound *(tzatzalum)*. Such conditions negate any positive merits soils may have from the standpoint of available nutrients. Thus it is that the type ranked lowest by informants (No. XIII) has the highest available nutrient content, that ranked third from the lowest (No. X) has the second highest, and that ranked next to the lowest (No. XII) has the fifth highest of all types sampled.

That soil texture is also an important consideration seems indisputable from the analysis. While those soil types ranked in first and second place by Chichipateños are high in available nutrients, that ranked in third place is decidedly low. However, the third ranking soil type, *samahi? li ch?och?*, is sandy, silty loam. In a similar manner, fourth, fifth, and sixth ranked soil types are soft and grainy, allowing for excellent percolation, and apparently giving good yields, though the latter two are also very low in available nutrients. In final analysis, the most desirable soil types would seem to be those that are black, porous, soft, and well drained. The least desirable would be those that are red, hard, and poorly drained.

All Chichipate soils are known as *kʔixnal chʔochʔ*, or warm soils, due to their location in the lowland tropics. Over half of them are reddish *(cak li chʔochʔ)* and, by Kekchi standards, hold limited promise for agriculture. According to local classification, about one-fourth of the soils possess a high clay content *(seb ru li chʔochʔ)*, as many as one-half are periodically covered by water *(haʔ ru)*, one-fourth are black *(kʔek li chʔochʔ)*, and a little more than one-tenth are designated as root-bound *(tzatzalum)*. The overlapping of these classifications seems obvious by the percentages involved.

Only isolated spots are rocky, and these are near the hills on the side of the community owned by the International Nickel Company. A distinction is made between soil with blue stones *(rax pec)*, white stones *(sak pec)*, and red stones *(cak li pec)*. In the words of one informant: "Blue and white stones keep cool, but the red stones get hot." The blue and white are considered good for wet season milpa; the red, poor. Soils choked with stones, if the stones are not red, are thought to be adequate and, in certain circumstances, even good for milpa. Neither are boulder strewn soils scorned, for, with digging sticks, certain problems associated with plow agriculture do not arise.

Types of Vegetation

Kekchi who migrate to the Izabal area rapidly acquire a minimal working knowledge of the complex forms of local plant life. Because these plants are looked upon as primary variables in the selection of productive milpa sites, the value of a particular piece of land is usually judged as much in terms of vegetative cover as in terms of soil type. Such cover not only reflects soil type and drainage, but seems to offer the principal basis for fertility, crops being maintained more by decaying vegetation than by the soil per se. Under highly leached conditions, the soil itself may serve as more of a holding agent than anything else (Geertz, 1963: 20-22).

The migrant Kekchi have brought with them the terms they use to designate principal types of vegetative cover. The most general ones are *guamil*, a Spanish derived term, which simply means "low secondary growth," and *guatal*, another Spanish derived term, which could be glossed as "higher secondary growth." In addition, there are a number of more specific Kekchi terms, though they are general enough to serve the need for classification as well in the lowlands as they do in the highlands (Table 7).

In speaking of ideal lands for preparing wet season milpa, Chichipateños emphasize those covered with high forest *(nīnk li q'uiche')*. The buttresses of many mature trees present the worker with very special problems. However, though clearing is difficult, burning is easy, and once large timber has been cut and burned, the soil is well-supplied with readily available nutrients in the ash. Weeding is usually not required during the first growing season on land just cleared from high forest.

TABLE 7
MAJOR TYPES OF VEGETATION

Kekchi term	English gloss
Rok cuaj	A field covered with small plants which appear in the first few months after harvest (literally, dried corn stalks)
Coc' pim	Low weeds
Pim ru	Thick, low weeds
Mas pim ru	Very thick brush
Ichaj ru	High grass, e.g., pasture grass
Coc' che' ru	Sapling stage of secondary growth
Q'uix ru	Thorny vegetation
Nīnki al c'al	High secondary growth
Mac'a' xpimal	Higher secondary growth, characterized by sparse undergrowth
Nīnk li q'uiche'	Mature forest growth*

*By the very nature of things, *nīnk li q'uiche'* is an indefinite category. It literally means "big the forest," and so cannot be used with the preciseness that Ursula Cowgill employs when she arbitrarily calls "climax" forest anything that has rested at least 25 years (U. Cowgill, 1962: 281).

In selecting a high forest site, no distinction is made between primary and secondary growth. This may be due to the fact that the Kekchi of Chichipate have lived in the region so few years as to be unaware of any lands cut over formerly, if, indeed, such lands exist. There certainly seems to be no awareness of the differential burning qualities or pest problems of primary over mature secondary growth. Simple observation confirmed informants' claims that milpas planted in places that were in mature forest the previous year are freer from weeds than those planted in lands where saplings *(coc' che')* or bush *(chamal pim)* had stood.

According to local theory, when mature forest growth is not available for milpa preparation, preference is given to those lands covered by tall though not mature forest. Next in line come lands covered by saplings, and finally those covered by low, ligneous growth. Herbaceous fallow is thought not to be conducive to successful milpa, and grasslands are looked upon as useless, though

TABLE 8
PLANT INDICATORS FOR SITE SELECTION

Wet Season Milpa		Dry Season Milpa	
Kekchi type	Botanical species	Kekchi type	Botanical species
Yaxte'	*Brunfelsia* sp.	*Quēnk' caballo*	*Stizolobium* sp.
Tz'uyuy	*Cochlospermum* sp.	*Tz'imaj*	*Momordica charantia* L.
Tz'ukl	*Pithecolobium* sp.	*K'erk*	*Heliconia latispatha* Benth.
Hu	*Virola guatemalensis* (Hemsl.) Warb.	*Yaxte'*	*Brunfelsia* sp.
Cuāchil	*Dialium guianense* (Aubl.) Sandw.	*Ak'l*	Moraceae
Muy	*Lucuma durdlandii* Standl.		
Pōk	*Spondias purpurea* L.		
Chahib	*Trema* sp.		
Puj	*Ochroma lagopus* Swartz		
Yaxhab	Unidentified		
Kinam	*Vatairea lundellii* (Standl.) Killip ex Record		

redeemable through the planting of legumes, especially *quēnk' caballo,* or velvet bean.

A whole series of specific plants, rather than plant associations, are believed to hold clues for the selection of a milpa site (Table 8). Some are used as diagnostic criteria for determining where to plant wet season milpa, while others are used for selecting parcels for dry season milpa. Plants indicative of good wet season land are those which have a relatively low tolerance for poor drainage conditions;

those indicative of good dry season land have a high tolerance for such conditions. Only one plant is considered promising for both types of cultivation, and this is *yaxteʔ* (*Brunfelsia* sp.).

For both dry and wet season milpa, land covered with soft-wooded timber is considered generally good, especially when the soil texture is loose and humid.[8] This soft-wood growth is exemplified by *hu* ((*Virola guatemalensis* (Hemsl.) Warb.)) and *puj* (*Ochroma lagopus* Swartz).

The proliferation of three types of plant life is thought to invalidate a piece of land as a milpa site. All of these types are grasses: *pachʔayaʔ* (*Paspalum conjugatum* Bergius), *cak pachʔayaʔ* ((*Axonopus compressus* (Swartz) Beauv.)), and *acʔ* (unidentified). The first two are stoloniferous. The third is a type of bush grass used for thatching.

Distance from Human Settlement

Except for a few house yard plants, few crops are sown in close proximity to houses in Chichipate. Practically all milpas are at least a fifteen-minute walk from the hamlet, and many are an hour's walk or more away. The constant search for more productive land accounts for this scattering to some extent, but the coercive forces of the two new landlords seem also to be powerful determinants. Perhaps the single most important factor, however, is the presence of hundreds of pigs in the hamlet. Since they act as scavengers, these animals are never penned in. Allowed to run loose, they raid and completely destroy any unfenced corn fields that are near the hamlet.

Roads and paths invariably take precedence over milpa planting. Major roads, such as the one running between Cobán and El Estor, cannot, of course, be used for cropping, even if one would wish to try. But even minor roads, such as the one leading from Chichipate to the Yuscarán *finca,* are considered inviolable. Footpaths, of course, run all over the region. Milpa can be planted to either side of these, and seldom is the path diverted for the sake of milpa unity. As people walk along these paths, they gather grass seeds on their clothing. Damage to growing crops results not only from direct tramping, therefore, but also from the sizeable mats of grass that spread on either side of the path. Yet land is still abundant enough

8. Soft woods are generally found in early succession, hard in later. For further discussion of this matter, cf. Budowski, 1963.

so that walkers are not forced to skirt fields. Given more demographic pressure, this laissez-faire attitude may well be expected to change.

Terrain

The half of Chichipate lying south of the El Estor-Cobán road consists of even land which gradually drops to the flood plain of the Polochic River. Most of this land floods in the rainy season and thus is of only marginal use for wet season milpa. To the north of the road the plain gradually slopes upward until it brusquely meets steep foothills, some of which intrude into the hamlet site itself. The portion of the plain lying north of the road offers the best sites for wet season milpa, and, until the dissolution of the Tinajas cooperative and the subsequent partitioning of the community (cf. p. 3), farmers tended to plant on the north side of the road in the wet season and on the south side in the dry season. Partitioning upset a neat ecological balance that had existed ever since the first colonists settled in Chichipate.

Lands that easily flood *(haʔ ru)* are considered unpromising, particularly for wet season milpa.[9] Flooding, in fact, accounted for more crop failure in 1965 than any other single factor. Beyond a general tendency to avoid depressions, however, there is little consideration given to terrain as such. Well-drained valley *(ru takʔa)* or level lands *(helho ru li chʔochʔ)* are highly desirable types. But sloping land, whether slightly sloping *(mero salso)*, moderately sloping *(mas salso* or *mero tzūl)*, or extremely steep *(mas tzūl)*, is often preferred to level land because of the greater ease milperos claim it offers for slashing and felling. An additional initial advantage may lie in the possibility that rapid exposure of the subsoil through erosion brings to the surface mineral nutrients previously leached.

In no case is terrain the only consideration for the selection of a milpa site. Soil color, soil texture, soil condition, and vegetative cover all loom as much more important. The slower replenishment of secondary growth on badly eroded slopes leads the Chichipateño

9. The reluctance to use *bajo* land seems widespread throughout the lowlands. In San José, Reina found that several milperos were very happy with their sites because they were not in a *bajo,* the soil was uniform and of the best type, and the location was in a wide space, high and open to sun and air, and on a slight slope, oriented so that the heaviest winds could cause little damage (Reina, 1967: 3).

to implicitly recognize the effects of the erosion process over time. This recognition, however, does not restrain him from cutting over the steepest of these slopes, nor from planting in an up and down direction rather than on the contour.

Irregular terrain is described by the Kekchi according to its specific features rather than in general terms. Digging stick technology makes the use of such terrain feasible, though depressions between small slopes where flooding is likely to occur are usually left to grow up in weeds and small shrubs.

The base of a hill or mountain is known as *xton tzūl* and the middle as *xyitoc tzūl.* Both the base and the middle are commonly used for milpa sites. A number of hills around Chichipate have rather level summits. These, known as *cocʔ takʔa* or little plains, may be used for milpa and, except for their relative inaccessability, are considered to be especially good sites for wet season cropping.[10] However, because of distance from the village, they are seldom used.

Religious Considerations

One of the effects of migration from Alta Verapaz to Chichipate seems to have been the lessening of supernatural sanctions. In the Cobán area, hills are commonly associated with various saints. At the time of site selection, should one have a dream about any of these hills, or any of the associated saints, he cannot make milpa there.

In Chichipate there is less respect for the saints. Most hills have not yet been given supernatural associations, and so people make milpa wherever they desire, so long as this desire does not run counter to the mandates of the present landowners.

The spirits of hills in general must be placated, however, at the time of boundary marking, planting, and harvest. Theoretically, milpa may not be made in a site if the spirits of the surrounding hills have not given permission. Yet the general attitude of the Chichipateños is pragmatic in the extreme. As one informant put it: "Everyone has to eat, so we make milpa anyplace we wish as long as we ask permission first. God would not have us die of hunger. Where could we go? After all, we are all children of God."

10. Reina, in speaking of site selection in San José, indicates that these small plains, along with their immediately adjacent slopes, constitute the core of milpa concentration for that community (Reina, 1967: 3).

Discussion for Site Selection

Milperos do not like to waste their time on land they consider unpromising. Site selection is often a difficult task, for there are relatively few large tracts of land that display entirely positive features. In the selection, older colonists of Chichipate have a distinct advantage over those who have come into the community more recently. They may have as many as ten to twenty *manzanas* (approximately 28 to 56 acres) in cleared-over lands, most of which are in various stages of secondary growth. With such extensive reserves, these men plan from three to five years ahead which plots they will submit to recultivation. Newer arrivals to the community, however, may not decide on a milpa site until January or February, i.e., three or four months before burning for wet season milpa should occur. As they walk through the bush, they constantly search for a piece of land that seems to meet best the various requisites of soil color and texture, vegetative cover, accessibility, and terrain. If they see a likely spot, they tell their friends, brothers, fathers, or fathers-in-law, and invite them to inspect the find. The first person who locates such a site opens a swath about a yard wide around its perimeter to indicate that it has been taken. Should the site have been formerly used by an individual still resident in the community, that person may appear and claim it as his own. At times such rival claims lead to open fights. At other times, the auxiliary mayor, a local resident appointed by authorities in El Estor, intervenes and decides who has prior claim. If the site is unusually large, the auxiliary mayor may opt to divide it into two parts. Only on rare occasions is the former occupant left without any rights whatsoever.

Generally, a milpero tries to have productive plots in several places at the same time. This custom of scattering was recorded by Bishop Landa for Yucatán in the early Colonial period,[11] and was attributed by him to a desire to insure a minimum harvest, no matter what the seasonal conditions. The ideal pattern dictates that once a person has cleared over land, he has primary access rights to it as long as he remains in the community, with the exception, of course, of intervening rights on the part of the absentee landowner. Rights to land can be passed on to a relative. Proper behavior demands that newcomers, if they cannot find adequate sites covered by high forest, go to older residents and beg for the loan of lands

11. Landa's words are: "*Siembran en muchas partes, por si una faltare supla la otra*" (Landa, 1938: 111).

which are not immediately needed. Such requests are usually directed to relatives, though occasionally to unrelated friends. Permission to use a site is given for but one season at a time, and recompense is purely in the form of social loyalty, including participation in work parties of the lender.

The fact that the absentee owners of Chichipate today prohibit indiscriminate cutting over of virgin timber means that in-migration may soon be stopped by the milperos themselves. The older residents of Chichipate claim to have loaned about as many milpa sites as they can afford to, and, as a result, some of the newer ones find access restrictions so tight that they must constantly supplement milpa activity with wage labor in order to maintain even a minimal level of subsistence. These wages are usually earned while working as day laborers in surrounding *fincas*. It is understandable that the ties these newcomers have to the community are loose indeed.

Timing

Kekchi migrants who have settled in Chichipate are sufficiently involved in the dominant European culture of Guatemala to follow the European calendar in determining the time for planting. Formal selection for wet season milpa sites usually takes place in February, though mental notes may be made for one-half to five years in advance. Slashing and burning must occur before the first rains begin *(habal kʔe)*, the best burns being obtained in the hot dry days toward the end of April and the beginning of May. A thorough burn is impossible once the first heavy rain has fallen, and rain is a certainty by the last weeks of May.

Phases of the moon determine the time for planting of a number of items, particularly those which are intercropped. Malanga, manioc, and sugarcane should be planted in the week immediately following the full moon, i.e., when the moon rises from 8:00 to 9:00 in the evening. Bananas and plantains are ideally planted when the moon is in the middle of the sky at 6:00 or 7:00 in the afternoon.

Ritual

There exist excellent descriptions of agricultural ritual for parts of the lowland Maya area. That presented by Redfield for Chan Kom is immensely more complex than anything which today can be observed in Chichipate (Redfield and Rojas, 1962: 127-47). The same may be said for Gann's descriptions of ceremonies reported in his work on the Maya of Yucatán and northern British Honduras (Gann, 1918: 40-48), and for Thompson's observations on the Kekchi of southern British Honduras (Thompson, 1930: 56-118). In all these works, the main ceremonies are described as occurring at the cutting of the bush, planting, ripening, and harvest. These are still the major ceremonial points in the agricultural cycle of Chichipate.

Simplification of ritual forms might be expected as normal for a small migrant population. The fact that Reina has recently reported the same phenomenon for San José, a long established lowland settlement, suggests, however, that secularization may be a growing trend for all lowland groups. In his article on the abandonment of *primicias,* he cites fear of performing poorly because of inadequate knowledge as a principal factor in the trend.[12] He likewise points out that though in the past San Joseños practiced private rituals in their fields to protect their corn from thieves, they have abandoned that practice as well. The reason given is that the ritual might endanger the lives of relatives and friends who might enter the milpas without proper permission (Reina, 1967: 16).

Fear of supernatural reprisals, should ritual not be carried out properly, was found to be ubiquitous in Chichipate. In the absence of *ah-men,* the traditional Maya part-time religious specialists, milperos were forced to rely upon what little they could remember from their fathers and grandfathers. Two young men in the community had memorized some Catholic canticles, and were called upon for saints' days celebrations and life crisis ceremonies. But these young men were next to worthless for the milpa cycle. Here the most respected were the aged of the community, and these were frequently consulted as to proper procedures for selecting, planting, caring for, and harvesting the milpa.

At the upper end of Chichipate, near the El Estor-Cobán road, stands a simple structure, open on three sides and thatched with palm. This building, the *ermita,* corresponds to the aboriginal Dolores hut, a structure which was set aside as combined temple and men's house (Thompson, 1938: 596). In aboriginal times, these structures housed the idols. Today, in Chichipate, the *ermita*

12. In the words of one informant, "It is preferable to take a minor damage, one death, rather than expose ourselves to much danger because of failure of participants to observe the things prescribed in the ritual. The *ah-men,* who are few nowadays, are afraid to undertake the responsibility of the *primicias,* and we cannot afford to lose them" (Reina, 1963: 223).

houses the images of Saint Dominic and Saint George. These stand on a table against the northeast wall, and are set off from the "nave" by a low picket fence. The ancient Maya made sacrifices to woods, very high and rough mountains, dangerous passes, cross-roads, and great whirlpools in rivers, believing that from these came everything needed in life. Sacrifices generally took the form of copal incense and turkey blood sprinkled on pieces of pine (Thompson, 1938: 594). Such practices still form the heart of ritual in Chichipate.

The second evening before selecting his wet season milpa site, each household head goes to the *ermita* to formally ask permission and blessing from various spirits. He sets the date for this visit some two or three days in advance. In order to carry out the proper ceremonies, he buys from two to twenty-five candles, each of which costs one cent. These offerings he supplements with copal that he has collected during summer forays through the bush.

As he enters the *ermita* with his wife, they step inside the picket enclosure and kneel. Then the milpero prays:

> Lord Jesus, I beg to come before you. I wish to ask permission. I am about to begin preparing my fields. The day after to-morrow. I want you to free my way, and to help me reach my fields, Lord Jesus Christ. Here I am to beg you.

At this point both the milpero and his wife place burning candles on the table before the saints. Then the copal brazier is lit, and the woman shakes the incense toward each of the saints. She turns, walks to the edge of the chapel, and shakes the brazier in the direction of the milpa site. She again offers incense to the saints, repeats her homage to the milpa site, and for a third time propitiates both saints and site.

Prayers are offered by the household head to all important hills and mountains of the region, up to sixteen in number. To them, as well as to the saints and the spirit of the site, the milpero lifts the petition:

> I beg for my milpa, for my business. I beg you to grant me my request, for I humble myself beneath your feet.

More incense is burned and, when it and the candles have been exhausted, the couple return to their home. There they retire and rest the entire next day. On the following morning, at 2:00 or 3:00

A.M., they rise and burn copal and candles before the saint of the house, often merely a colored print taken from a religious calendar or magazine. On making these offerings, they repeat the prayers they had uttered a day and a half before in the chapel. A chicken is then brought in, and its neck wrung by hand. The bird is opened and washed in cold water, and the bloody water is saved in two empty wine bottles.

At about 5:00 A.M. the milpero and his wife go to their milpa site. In its very center he places a small cross made from sapling branches. This cross (*kā cuaʔ*—god cross) is built to stand about 1 foot high. The milpero takes one of the bottles of bloody water and shakes its contents around the foot of the cross. The contents of the other bottle he scatters toward all sides of the site. A candle is then lit and placed at the foot of the cross. Finally, the brazier filled with burning copal is shaken, by the household head rather than by his wife, toward the corners of the milpa site.

Accompanying prayers vary with the individual. Their content, however, centers around the need for protection from natural and supernatural threats. One is as follows:

> In the name of God the Father, Son, and Holy Spirit. Be with me, free me, give me your blessing. Help me in my struggle to make milpa.
> Keep all serpents from me, all animals. Keep me from harm. May no trees fall upon my body. Free me. Be with me through the Holy Spirit. Amen.

Since ritual is but a small part of the marking process, the couple and their helpers usually stay at the milpa site until nearly noon. Upon arriving home, they give food to the household saint. This is customarily boiled chicken and broth, tortillas, tamales, and whipped bitter chocolate. The offerings are left in gourd dishes either on a small table or on the ground in front of the saint's image. They are presented together with burning candles and copal incense. As was done in the chapel two days before, the incense brazier is finally shaken alternately before the saint and in the direction of the milpa site three times, and this concludes the ceremony for site selection.

Chichipateños tell of divination practices which are used in conjunction with site selection in Alta Verapaz, but apparently these were lost in the move to the lowland tropics.

Measuring and Marking

A man and his wife may be accompanied by one or more male helpers when they go into the bush to mark their milpa site. These helpers are commonly kinsmen. The first thing done upon arrival at the site is to make a path about 1 foot wide to the center of the milpa, and there to make a small clearing around the ritual cross *(kā cuaˀ)*. An arm's length of vegetation is slashed on all sides of the cross, and the trash is pushed back into the surrounding bush so as to leave the ground around the cross completely bare. Offerings of blood, incense, and candles are given, and prayers are said. Only then does the actual measurement of the site proceed.[13] Reflecting its role in the ritual protection of the milpa, the clearing around the cross is known as the heart of the milpa, *rām li cˀal—rām* coming from the Spanish *ánima,* or soul.

The basis for measurement is the *tarea,* or task. This term refers to the amount of land a mature male may realistically expect to slash in one working day; how extensive this is depends on the density of vegetation with which he has to cope. A wet season *tarea* in Chichipate comes to 30 *varas* square, though in Murciélagos, the Hempstead estate on Lake Izabal, it comes to only 25.[14] The *vara* is also variable. Theoretically averaging 2.8 feet, it comes to about 35 inches (89.4 cm) in Chichipate. It is determined by stretching, with the arm fully extended, a string from a man's middle finger to the center of his breastbone.

Measuring is carried out by using a rope or cord. A pair of stakes from 3 to 5 feet high is placed at the end of each 30 *varas*

13. Redfield described the marking ritual of Chan Kom as follows: "If the field is being cleared for the first harvest *(chăcben),* the agriculturalist is ordinarily careful to measure off for his milpa no more land than he is sure he will be able to clear. He cuts a narrow swathe around the proposed field, and by this act serves notice on the *kuilob kaaxob* that he proposes to make use of so much of the 'virgin bush' *(zuhuy kaax).* Then when the trees ask the *kuilob kaaxob* if they may not do injury to the man when he begins to cut them down, the deities say no and explain to the trees that they are old and that it is good for them to be cut so that new growth may come up" (Redfield and Rojas, 1962: 132).

14. Bishop Landa described a measure similar to the *tarea* in colonial Yucatán when he wrote his *Relación.* The unit appears to have been considerably smaller than that used in Chichipate today: *"Suelen, de costumbre, sembrar para cada casado con su mujer medida de 400 pies lo cual llaman hum uinic, medida con vara de 20 pies, 20 en ancho y 20 en largo"* (Landa, 1938: 111).

(i.e., at each *tarea* division marking). The stakes are slashed obliquely across the top and set about 3 inches apart. From 1 foot to 18 inches of growth is cleared around them, and a path is then made to the next pair of stakes.

Milpas larger than a single *tarea* are customarily calculated in terms of *manzanas,* a unit made up of 16 *tareas* (2.81 acres in the English system). Not all milpas are square or four-sided *(cāxucūt).* They may also be long and narrow *(caˀchˀin ru),* wide *(nim ru),* curved *(kˀotkˀo),* or circular *(sursu).* Irregular forms are subsumed under one of these five basic categories.

From the standpoint of size, a milpa under 1 *manzana* is considered small *(chˀina cˀal)*; from 1 to 3 *manzanas,* medium-sized *(mero yuˀyu li cˀal)*; and above 3 *manzanas,* large *(nimla cˀal).*

Division of labor for the measuring and marking process is relatively simple. If more than two men are involved, the older, especially if he is an affinal relative of the household head, conducts the ritual. Men alone slash, measure, and mark. Women say prayers and burn incense, but do little more.

Cutting

When highlanders move into Chichipate they are forced to learn different techniques of field preparation from those they knew. First of all, the omnipresent highland hoe is seldom needed in lowland cultivation; second, the lowlands permit two crops of maize per year whereas in the highlands one is normal; third, secondary tropical moist forest characteristics require coping with very dense low vegetation, as well as with heavily buttressed trees. Clearing such cover consists of three distinct stages: the cutting and slashing of herbaceous and sapling growth *(cˀalec),* the felling of heavy arboreal growth *(tˀanoc),* and the burning of the resulting slash *(cˀatoc).*[15]

Slashing (cˀalec)

Slashing must begin several months before the planting season. The first milpa cut in 1965 was begun on February 1, and the last was finished by May 1. The early part of March is generally con-

15. With the exception of the substitution of metal for stone tools, clearing techniques seem to have changed little in the lowland Maya area since colonial days. Landa's *Relación* puts it simply: *"En labrar la tierra no hacen sino coger la basura y quemarla para después sembrar, y desde mediados de enero hasta abril labran y entonces con las lluvias siembran"* (Landa, 1938: 111).

sidered to be the latest possible time for selecting and marking a site.

The fixing of a date for beginning slashing correlates directly with the size of growth that must be cleared. Mature forest growth *(nīnk li q'uiche')* must be attacked by February 15, if for no other reason than that a *manzana* of such growth will have from twenty to thirty buttressed trees. Cutting of relatively high secondary growth *(nīnki al c'al)* can be postponed no later than March. Clearing of sapling growth *(coc' che' ru)* must begin by April, and only low herbaceous growth *(coc' pim li ru)* can be left for the first of May, i.e., approximately two weeks before the latest time for planting wet season milpa.

Milperos may slash during any phase of the moon. For three days before initiating slashing, however, they cannot sleep with women nor eat fruit. Before beginning to cut, prayers of the same type that were used for the selection ritual are said in the chapel and in front of the cross that lies at the milpa's heart. No offerings need accompany these prayers.

Clearing of sloping land is considered easier than clearing of level land. The preferred method is to go from the bottom up *(taksīnc)*, for if one should proceed from the top down the slashed material would have no place to fall, and spreading would be quite difficult. Ease in slashing a hillside is countered, however, by increased difficulty with control of fire at the time of burning. The measure usually taken to ameliorate this difficulty is to fire the hillside from the top down. Soil quality, drainage, and vegetation, rather than slope, thus remain the primary criteria for the selection of milpa sites.

Small bushes, weeds, vines, and other herbaceous growth are the first things attacked. For this broad clearing work, a long, straight machete *(ahin ch'ich')* with a blade of 27 inches (68.58 cm) and handle of about 5½ inches (13.97 cm) is preferred. The work is done with wide, sweeping strokes near the ground, and there is little concern for obtaining an even cut. Milperos believe that fire will level things off anyway. Termites *(c'ams)* are known to be a problem in slash that sits for more than six weeks, though they are most active after the rains begin; by that time most milpas have already been burned. There seems to be little intentional saving of low vegetation for a later slashing date in order to cope with this problem; rather the ideal is to let no more than one month pass between final clearing and the initiation of burning.

Usually all small growth is cut at one time, bringing down everything that can be easily cut with a machete. Even saplings are felled in this manner, their stumps being left from a foot to a foot and a half above the ground. Leaving such stumps makes cutting easier, and some Kekchi say that it also promotes sprouting for quicker regeneration of second growth. Once the sapling has fallen, its branches are lopped off with the machete and tossed around the milpa so that the surface will be relatively level. In clearing sapling-type growth, the short curved machete *(colima chʔichʔ)* is preferred. It has a blade about 23 inches (58.42 cm) long and a 5-inch (12.7 cm) handle.

When a large tree (more than 27 inches in diameter) is felled, it is hoped that smaller trees will be taken in its path, though preliminary notching of the smaller growth is seldom practiced.[16] For heavy timber to be most effective in clearing away the growth that lies in its path, herbaceous material should first be slashed away from around smaller trees of the area. It is surprising how devastating the felling of a forest giant can be even without such preparation, however. In 1965 one man felled large trees onto another's milpa before the latter could even begin to clear out underbrush. The second individual simply left the area as he found it; he did not even bother to lop off the branches of the largest trees. Yet his fire swept the field so clean that the site later produced some of the best maize in the community.

When the milpa floor has been completely slashed, taller palms are felled with an axe. The fronds of smaller palms that are within reach are sheared off with a machete. The trunks are left intact, however, in the hope that these trees will survive the fire. The thatching palm ((*Orbignya cohune* (Mart.) Dahlgren ex Standl.)) is not only a very serviceable plant for the milpero; it is protected by Guatemalan law.

Spreading

In Alṭa Verapaz, the ideal in clearing is to cover all the ground evenly with slash. Unless this is done, the burn is imperfect, the soil is left subject to erosion, and the milpa crop becomes uneven and poor. With the much heavier vegetation of the Chichipate region, however, less care is needed. Indeed, where there has been a dense

16. In his report on Hanunóo agriculture, Harold Conklin has given a detailed account of just such preliminary notching. In this respect, as well as in many others, the pioneer agriculture of Chichipate is far less complex than the well-established system practiced by the Hanunóo (cf. Conklin, 1957).

stand of vegetation, spots of bare earth may be left throughout the milpa in the belief that the intensity of the fire in the rest of the field will produce sufficient ash to cover them anyway. Only when dealing with spotty low or medium fallow is any effort made to spread herbaceous growth, and this is done with a stick.

What spreading does take place is carried out simultaneously with slashing. The general procedure is for the worker to slash with a downward, diagonal stroke, holding the machete in his right hand. With his other hand, he steadies the plant. Herbaceous and small ligneous growth are cut together. Only large trees are left for felling at a later date.

Land is still abundant enough in Chichipate to permit relatively casual clearing. The general attitude prevails that if a plot burns well, fine; if not, it is best to leave well enough alone.

Protective Clearing

Protective clearing has probably been practiced in the Maya lowlands for many centuries. Lundell, for example, found that every southern lowland ruin he visited had abundant stands of breadnut (*Brosimum alicastrum* Swartz). This, he felt, was no coincidence, for elsewhere in the tropical forest these trees are rare (Lundell, 1933: 72).

Today in Chichipate, occasional firebreaks *(makʔoc)* are made around trees thought worthy of saving. To protect trees in this manner, the firebreaks are usually made early, preferably in January. Not only is low trash cleared from around their bases, but large, nearby trees are cut in such a manner as to fall away from the ones to be preserved. Only first quality, vigorous growth is so protected, and then only a few species such as copal *(pom cheʔ)*, mahogany *(sutzʔujl)*, cedar *(rutzutz)*, zapote *(saltul)*, rubber *(quicʔ cheʔ)*, chicle *(chiquibul)*, and young, first-rate thatching palm *(mocōch)*. At the time of burning, the fire is started at the edge of the clearing that has been made around these trees, so that it will burn away from rather than toward them.

Milpa sites may contain a number of other useful plants which are given no special protection whatsoever. Balsa wood *(puj)* constitutes the hamlet's primary building material; however, when green, it is extremely difficult to handle and peel. Most milperos prefer to fell it along with other trees in their plot and to risk the possibility of the fire consuming it entirely. The amount of balsa timber left more or less intact after burning is still sufficient to meet building needs, and it is easier to drag the logs out of the field once underbrush has been eliminated. Scorching, furthermore, eliminates the problem of peeling the green logs.

A number of food plants may still live on in a milpa site after it has been left fallow and neglected for many years. Of these, sweet potato, malanga, plantain, chili, and manioc seem fairly resistant to the effects of fire and often sprout anew once the rains have begun. They are given no protection whatsoever. Other food crops, if present in the site, are destroyed through either the slashing or the burning process. The abundance of the natural environment would appear to make their protection more costly than it is worth.

Milpa Labor

In the process of slashing, most men work by themselves or in very small labor groups. These groups generally correspond to the basic kin unit which shares a common storehouse and lives either in the same or in contiguous houses. The most common grouping of this kind is composed of a father and his son or sons. Second in order of frequency are groups composed of a man and his son-in-law, the latter cooperating because of his responsibilities for post-marital bride service. Occasionally brothers or brothers-in-law may form a labor group and, if so, share granaries and residences.

Other labor arrangements include reciprocity (*sachoc rekaj li cutan*—"to owe the day") and wage labor (*tojbil*—"paid" or *xtojbal li cutan*—"payment for the day"). Both of these arrangements may be established among all types of people, though they exist most frequently among affinal relatives and long-established friends. If reciprocal labor is recruited, an equal return is expected at some near future date. Wages for hired help are standard: 50 cents or 100 ears of corn per man-day of labor.[17]

Though a businesslike attitude prevails during cutting, felling, and burning, a festive air is evident during seed time. Then groups of ten to fifteen individuals organize for even small milpas, and intersperse planting labors with feasting on pig, turkey, and chicken.

17. Bishop Landa has given us a description of reciprocal labor for colonial Yucatán: "*Que los indios tienen la buena costumbre de ayudarse unos a otros en todos sus trabajos. En tiempo de sus sementeras, los que no tienen gente suya para hacerlas, júntanse de 20 en 20 o más o menos, y hacen todos juntos por su medida y tasa la labor de todos y no la dejan hasta cumplir con todos*" (Landa, 1938: 111).

The host provides the food and also assumes the responsibility of reciprocating when his helpers initiate their own planting endeavors. The ideal in planting is to complete an entire milpa in a single day.

For harvest, labor exchange is rare. On an emergency basis, one may hire five to six (rarely more than seven or eight) men, the advantage being to complete one's harvest before kernels begin to loosen on the ears and before rats and other pests make major depredations on the crop. Such group endeavor is usually limited to a single day, however, because of the need each person has to take care of his own milpa. After the first day of group activity, if one is to continue harvesting, he must do so alone.

For all milpa effort, boundary lines are strictly drawn. The only cases in which there are no distinguishing marks are those where father and son or son-in-law make their milpas side by side *(lakʔ-loqueb)*. But since these two also share a single granary, and divide up the proceeds from any sale of maize, they may be considered to form a single milpa group. This type of sharing arrangement has been noted in the vicinity of Lake Petén by Reina: "Both the smaller and larger milpas may be worked by a man alone or with the aid of sons or brothers. Interestingly enough, when larger milpas located at equal distances from the town are visited daily instead of by the week, or by the week instead of on a *temporada* basis, milperos always employ the aid of sons or brothers" (Reina, 1967: 11).

No evidence was collected in Chichipate on the relationship Reina notes between the distance the milpa lies from the hamlet and the establishment of this type of arrangement. However, in both San José and Chichipate such close cooperation seems never to develop among distant relatives, and certainly never among mere friends.

It may be said, then, that there are four major types of labor in the Chichipate milpa cycle: (1) a man working alone *(mac-ʔanjelac xjunes)*; (2) close cooperation by members of the same labor group who farm milpas side by side *(lakʔloqueb)*; (3) reciprocal labor, related as well as unrelated to feasting *(sachoc recaj li cutan)*; and (4) paid labor *(xtojbal li cutan)*. Of these four types, the most common by far is that of a man working alone. Except for planting activity, milpa labor groups, when they exist, tend to be composed of the male members of a single household, and these, in turn, tend all to belong to a single nuclear family.

Time and Labor in the Clearing Process

Slashing, felling, and burning the heavy tropical growth of Chichipate is a time-consuming process. For cutting over a 1-*manzana* field (2.8 acres) covered with saplings, milperos worked an average of 113 hours in 1965. The clearing of dense, low undergrowth requires fewer man-hours. On the average, milperos spent 84 hours in clearing 1 *manzana* of such growth in 1965. Clearing of climax forest requires the greatest expenditure of all. Milperos averaged 368 hours per *manzana* of *nīnk li qʔuicheʔ* in 1965. At the low end of the scale then, a man working alone needs about eight days for clearing a *manzana* of land. At the high end, he needs about thirty-six.

Intervening factors prevent men from working consecutively until their milpas are entirely cleared. On the south side of Chichipate, the lure of paid *finca* labor draws the men away periodically during the clearing season. And all over the community the concept of the week *(xamān)* and Sunday *(domingo)* exists. Though there are no special religious observances on Sunday, on that day clearing activity is suspended. Fuel is stored, houses are cleaned, and men go off to hunt and fish. The only agricultural activities generally practiced on Sundays are the more urgent ones of seeding and harvesting; burning and slashing are, at least ideally, limited to weekdays. Nor are all of these ordinarily available. Even if a man takes on no wage or reciprocal labor, three to four days a month must be taken off to weed intercropped items that are carryovers from the previous season, and a not inconsiderable number of days are lost to illness, in particular malaria.

Care of Slashing Equipment

Both curved *(colima chʔīchʔ)* and straight *(ahin chʔīchʔ)* machetes are commonly used by left- as well as right-handed men. Each worker carries with him to the milpas a small hand file to sharpen his instrument. He stops about once an hour, places the machete in his left hand, and strokes the file outward from the body until both faces of the blade reach razor sharpness. The procedure is generally carried out more easily with the heavier and straighter than with the lighter and more curved machetes. Light instruments tend to be made of harder metal and are thought to ruin files quickly.

Machetes are commercially manufactured. They sell for from Q2.00 to Q2.50 in one of Chichipate's two stores. If the instrument breaks off near the tip, it may be repaired by filing. If it breaks off across the blade, however, it must be sent to Panzós, for there are no forging facilities in Chichipate. Forging charges come to about one quetzal. Should the blade break off near the handle, the handle is simply reset lower on the blade.

Children use small machetes with mahogany handles. These sell for only sixty cents. Both children and adults, when they slash, keep the beveled side of their instrument down, facing the ground.

Tools for Felling

Once the underbrush has been cleared away, preparations begin for felling tall growth. Store-bought axes *(māl)* are used for this purpose. Though they all have but a single bit, they fall into two basic types: *bolbo māl* is an axe with a round eye, and *patzʾpo māl* an axe with a square eye.

Only the axe head is purchased, and this comes from stores in Panzós, El Estor, Cobán, or from a little store located across the road from the Chichipate school and owned by a Kekchi man. The handle is carved from local hardwoods by the milperos themselves. The smaller end of the handle is fitted through the eye of the head from the top and is always secured with a tight fitting wooden wedge. In spite of such care, accidents still occasionally occur, when the axe head goes flying off its handle.

To fell any hardwoods with trunks larger than 3 to 4 inches in diameter, the milpero must use his axe. Large, buttressed trees, characteristic of climax and old second growth *(nīnk li qʾuicheʾ)*, take up to two man-days to fell. As a result, unless they have a broad canopy, large trees are allowed to stand. A full *manzana* of virgin land may have from twenty to thirty large, buttressed trees, of which as few as six may be felled. The rest, having but light canopies, are left to be seared by fire; this is usually sufficient to kill the tree.[18] Milpas are dotted by the remains of such standing, dead giants. Rot and termites quickly take their toll, however, so that by the end of the rainy season these large trees have begun to fall of their own accord.

Because felling proceeds better when several men are working

18. The practice of leaving some of the large trees has been reported by Redfield for Yucatán: "Everything is cut down except a few of the largest trees; these are left 'because a little shade is good for the growing corn in a time of drought, and because too much ash would result if they should be burned' " (Redfield and Rojas, 1962: 43).

together, most Chichipate milperos keep enough axes to supply a work party of three or four men. As the instruments grow dull, they sharpen them anew, using the same files that they keep for sharpening machetes.

Felling (tˀanoc li cheˀ)

Straight-trunked trees with little above-ground root development present no real problem for felling. But heavily buttressed specimens, common in tropical woodlands, offer the greatest challenge and excitement of the whole clearing process. Many of the buttresses (*xeˀ*—"root") rise to over 10 feet above the ground. Coping with them, particularly when the trunk is leaning, requires highly developed skill.

Buttresses tend to be so thin and flat that the widest and smoothest of them are sometimes saved for tabletops. The worker must get above them to effectively attack the bole of the tree. Only one type of scaffolding is used for this purpose, no matter how high the buttresses reach. This is an extremely crude apparatus, built by setting two long poles into the ground and leaning them against the trunk of the tree. Against these are lashed two other poles to provide a forked support for a platform. On top of this rest the two poles that constitute the platform itself. The whole is held securely with vines.

Scaffolds (*xtaklebal*—"instrument for going up") are built in such a way that the platform lies some 3 feet below the spot where the worker plans to begin chopping. They may rise as high as 9 feet above the ground.

The decision as to the direction in which to fell the tree is a critical one for which numerous factors come into play. The first side attacked is that toward which the tree is intended to fall (*xnatˀ bal* —"landing place"). The felling notches on this side, known as *nim xsaˀ xyocˀlal* ("the big cut"), are made deeper than those on the other side (*cˀachˀin xsaˀ xyocˀlal*—"the small cut"). In general the easiest technique is to fell the tree in the direction that it is leaning *(arin taˀilok)*. However, this is not always practicable. Entangling vines, neighboring plots, nearby paths, and vegetation that the milpero wants to preserve may force him to decide to bring the tree down in a different direction. In this case, he constructs his scaffold on the side opposite to the direction in which the tree is leaning, and, if he is right-handed, places the *nim xsaˀ xyocˀlal* to the right; if he is left-handed, it goes to the left.

The most common practice is for two men to work at the same time. These are often joined by another pair, with whom they take turns. If there is a strong wind, it is hard to predict whether the

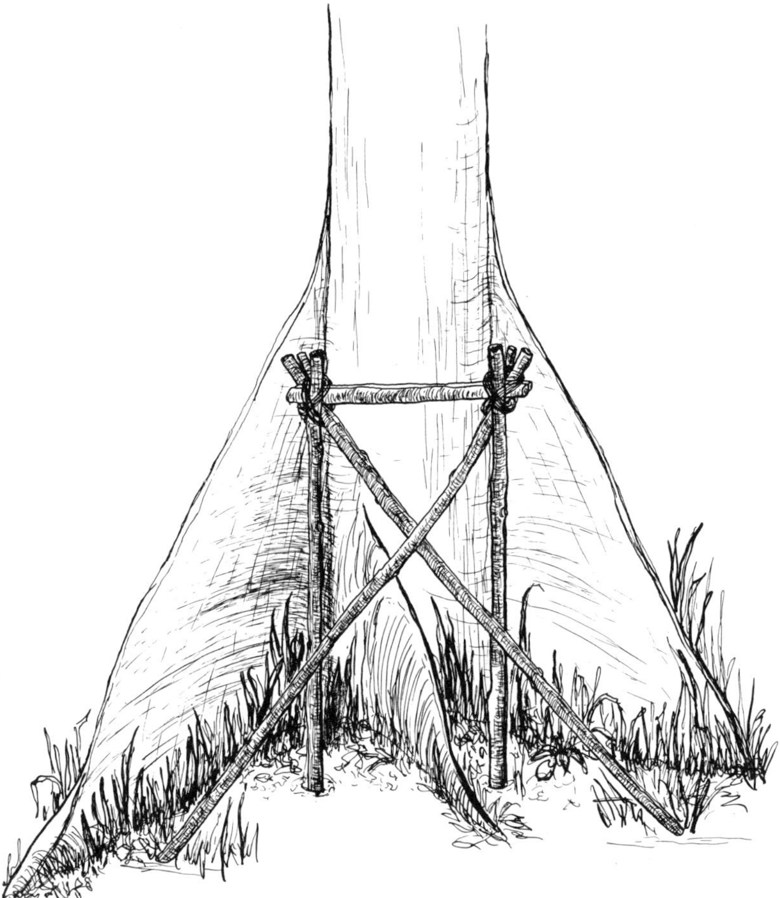

Scaffolding apparatus constructed by milperos to reach the bole of the tree.

tree will fall on the side of the big or the small cut, but it rarely falls in the direction of the workmen. As a precautionary measure, however, as soon as a crack is heard, the men run quickly away and shout *"cuan aran li cheʔ"* ("there's the tree"), or *"tʔanocʔ re li cheʔ"* ("the tree is falling"), as a warning to anyone else in the vicinity.

Accidents do occasionally occur. The three most common ones result from inadequate shaping of the cuts.

1. *Xmich?e·li che?*— "The tree is uprooted" because the large cut was made too deeply and broadly for the shallow cut.

2. *Xjol c?oc chiru lix tōnal*—"The trunk jumps back" because: (a) it is caught up with entangling vines, (b) it falls into the crotch of another tree, (c) the large cut has been made too low on the trunk, or (d) a combination of these factors.

3. *Xjache? lix tōnal*—"The trunk splits" because the large cut has been made too high. As the tree falls a diagonal split opens up between the large cut and the small cut, and it is impossible to predict the direction in which the tree will fall. This is one of the most dangerous of the abnormal falls.

Still another type of accident that terrifies milperos occurs when venomous serpents drop out of trees where they have been nesting.[19] In just the few years that they have lived in Chichipate, the milperos have witnessed several deaths from this cause alone. One who survived this experience is still badly crippled.

Occasionally cuts are made so badly that the tree fails to fall. This occurs especially when the small cut has been made too deep. *"Inc?a? us lix xyoc?bal"* say the workers—"The cut was no good."

Though entangling vines may result in a dangerous and uncontrolled fall, they are seldom cut away. By leaving them, much of the work of clearing out surrounding growth is eliminated, for large, vine-covered trees clear a wide, clean swath as they fall.

The scaffolding is not dismantled when a fall is made. If not destroyed by kickback or by the action of entangling vines, it is simply abandoned. A rickety structure to begin with and built from fragile soft woods (usually *hu* or *xc?ot tzic*), it is believed not worth bothering about.

Trimming and Pollarding

An example of the relatively low level of sophistication present in Chichipate agriculture is the fact that trimming and pollarding are

19. Several species of *Bothrops* found in the northeastern Guatemalan lowlands are arboreal: *B. undulatus* (Jan.), *B. bicolor* Bocourt, *B. nigroviridis aurifer* Salvin, and *B. schlegeli* Berthold. Perhaps the most likely which could be a common threat in Chichipate is *B. schlegeli* Berthold. Of this species, Dunn (1944: 216) has written: *"Su cola es prensil y la usa frecuentemente para descolgarse por las ramas de los arbustos."* Smith (1945) and Duellman (1963) may be consulted for additional information on arboreal venomous serpents of the area.

practically never practiced. Unlike Conklin's Hanunóo who were quite aware of the benefits in gradually spreading canopies during the growing season (Conklin, 1957: 62), when the milperos of Chichipate clear the land, their major interest is limited only to the desire for removal of obtrusive vegetation. Among the larger specimens, only those trees with broad, heavy canopies are removed. Others are left to stand and be burned. No effort is expended in attempting to preserve saplings, either for their intrinsic value or as climbing poles. Beans and sweet potatoes are about the only two cultigens which would benefit from climbing, and maize stalks provide all the support that is needed for them.

Once slashing and felling have been completed, a lull appears in the activity of the hamlet. *"Xinchoy rakʔinquil lincʔal"* says the milpero: "I have finished cleaning my milpa." He lies in his hammock, plays his marimba, gossips with his neighbors, makes a trip to El Estor, Panzós, or Cobán to purchase seed, works on a neighboring estate, hunts, fishes, builds fences, or collects chicle.

Burning

In the words of Carl Sauer, "The mastery of the forest by man requires no axe" (Sauer, 1958: 108). Fire may be used to fell a tree, cut it to desired lengths, and even crudely shape it. The thin-barked and hence poorly insulated trees of the humid tropics, particularly when young, are easily killed by burning. Extensive use of fire is frequent in those humid tropical lowlands where, as in Chichipate, a long dry season alternates with a relatively heavy rainy season. Such regions are very extensive and, regarding the tropical lowlands, the most heavily populated (Budowski, 1956). Since burning selects for certain plant species over others, most tropical lowland flora today is probably what it is because of periodic man-made fires (Denevan, 1961: 300).

The Chichipate milpero, as does his counterpart in lowland tropical slash and burn systems across the world, seeks five principal results from the firing process. First, he wishes to clear away the slash that litters his field. Given the simplicity of his tools, once he has cut low growth and felled buttressed trees, burning is the only means at his disposal. Second, he wants to loosen the soil so that dibble planting may be easier. Third, he wants to lower the rate of regrowth of herbaceous and ligneous plants. Fourth, he wants to reduce the number of animal and insect pests. And fifth, he desires to cover the ground with a fairly even layer of ash and thus provide readily accessible nutrients for his growing crops. Though he has no technical understanding of the nutrient needs of plants for which the ash may provide, he does clearly recognize that there is a direct correlation between the fertility of the soil and the depth of the ash layer.

Ash is broken down into two basic types by Chichipateños: white and black. The black is preferred by far because, in the words of one milpero, "it enters the soil." In contrast, white ash tends to form a dry, spongy layer on top of the soil. Ursula Cowgill, observing the same preference for dark ashes in the Petén, has pointed out that since organic matter, nitrogen, phosphorus, and exchangeable calcium all decrease with burning, their decrease would be greater for hotter fires than for cooler ones. In the Petén, as well as in Chichipate, fires used to clear fields generally leave black to gray ashes, indicating lower temperatures than those at which white ashes are produced (U. Cowgill, 1962: 282). Parts of Chichipate contain latosols which, when exposed to very high temperatures, develop a bright orange color from the oxidation of iron compounds; if this color penetrates very deeply, the soil is considered useless for milpa.

Several plant types are specifically named as conducive to producing one or the other of the ash deposits. The eye of the palm *(mocōch)*, for example, is believed to yield black ash. Tamarind *(cuăchil)* yields white ash. Heavy stumps that burn for days are known to be particularly productive of white ash and may leave deposits as much as 10 inches thick. Maize planted in such spots, if it germinates at all, withers before it ever reaches maturity.

Preparatory Activities

In order to obtain an adequate burn there should be a thick layer of debris distributed more or less evenly over the entire milpa. This debris *(cʔalem)*, the result of slashing activity *(cʔalec)*, varies from 1 foot in depth for low growth to 3 feet or more in depth for high, dense growth. It is imperative that most of this deposit be laid down by the middle of March, for burning usually commences the third week in April.

Perhaps due to the fact that Chichipate has been settled for such a short period of time, relatively little care is taken to secure an even thickness of debris. Where vegetative cover is poor, a man may at the time of slashing take a long stick and casually level the deposit; and once a tree has been felled, the branches may be lopped off and tossed around the milpa. This, however, constitutes the maximum of the rearranging process.

With hot, dry weather, debris turns brown in about one week, the first indication that it will soon be ready for firing. When it has completely shriveled, after four to five weeks, the firing time has

Drying slashed and felled material in preparation for burning.

arrived. If left for more than six weeks, debris begins to suffer from infestations of termites *(cʔāms)*. Most slashing is therefore completed approximately a month before anticipated firing time.

Firebreaks

Three basic types of protective clearing are practiced. The first is carried out at the beginning of the slashing process and is aimed at protecting semidomesticates such as copal, mahogany, cedar, chicle, and rubber. All trash is removed in a broad circle around the tree in question and the ground is left bare. When other trees are felled in the immediate vicinity, they are cut in such a manner as to fall away from the one to be preserved.

The second type of protective clearing may be carried out either at the finish of the slashing and felling activities or on the day of firing itself. This consists of the clearing of three to four yards on those sides of the milpa that are surrounded by low, secondary growth. All vegetative cover is removed until the earth remains bare, and the resulting debris is tossed back on the milpa. One exception to the need for such a firebreak comes when the milpa is surrounded by tall timber. In such a case, only underbrush is cleared away. When climax forest surrounds a milpa, no firebreaks are made at all. Theoretically, the fire will not penetrate such growth. In actual fact, however, it occasionally does, though it has never been known to get completely out of control in a climax situation.[20] A second exception to the clearing of firebreaks occurs when one milpa borders on another. Here it is of no importance if the fire spreads, and so firebreaks are considered superfluous.

Protection of the hamlet is the third consideration in constructing firebreaks. Though all milpa fields lie a brisk ten-minute walk or more from Chichipate, and the hamlet is surrounded by a ring of mature forest, sparks have been known to ignite houses in the past. Furthermore, the ring of mature forest has proved hospitable to marauding animals such as mountain lions, and so the women of the hamlet have recently been exerting pressure on their husbands to have it cleared away. In 1965, men went out at 8:00 A.M. on the day of burning and opened up a 30- to 40-foot firebreak in back of their houses, leaving the large trees, but removing all undergrowth.

One of the reasons firebreaks are used under only certain circumstances appears to be the fact that so much cultivation is carried out on the Polochic plain. Informants claim that fire is much easier to control on level than on sloping land. On a hillside, firebreaks are placed on all sides of the milpa and tend to be broader than those on flat land.

Timing

Though there is one main firing day for the entire community, individual burning occurs over a two- to five-week period. It begins some three days after the first thunder of the year has been heard in the hills.

Ideal burning conditions are those in which a bright, hot sun

20. The same has been noted by Steggerda in Yucatán. He writes: "In the Chichén Itzá area, fires die out at the edge of the cut bush and uncontrolled forest fires seldom occur . . ." (Steggerda, 1941: 95).

sears the ground, and wind is either absent or very gentle.[21] Noon is the best time of day for firing, for then the sun is at its hottest, and the gradually increasing afternoon breezes make for thoroughness of the burn.[22] The best burns occur when the wind comes from the southeast, the prevailing direction for winds up to the month of July.

If burning is not carried out at the proper time, the milpero encounters difficulties in cropping. In 1965, one man procrastinated until after the rains had begun, at which time he had severe problems in attempting to obtain a thorough burn. Some parts of his milpa never could be used.

Though milperos are very concerned about starting to burn before the beginning of the rains, firing milpas too soon can also cause problems. Planting may be so early that a good portion of the seed is lost, or weeds may get a head start on the crops and finally choke them out. The situation calls for careful reading of weather signs and swift action at the appropriate time. Even a successful burn does not guarantee a successful crop. Some milpas that did burn well in 1965 later flooded so badly that the yield came to less than 8 *quintales* per *manzana*.

Firing Techniques

Because of the danger uncontrolled fire represents to valuable woodlands, the Guatemalan government has set up a forestry service which, among other things, supervises firing for slash and burn agriculture and collects a 25-cent fee for each milpa that is so cleared. The agent in charge of the Chichipate area lives in Puerto Barrios, a good two days' journey away. Consequently, he seldom appears. In his stead, the municipal government of El Estor authorizes Chichipate's outgoing auxiliary mayors, just before they leave office, to appoint a group of *comisionados* to supervise burning. These men are usually chosen from among the younger members of the community, for agility is an asset in controlling fire. They may not refuse the responsibility; if they try, the force of Guatemalan law intervenes. Chichipate, in 1965, had twenty-four of these officials.

21. Chichipate practice, in this respect, conforms to that of the Tajín Totonac (Kelly and Palerm, 1950: 110).

22. Relatively little has been written about timing of the burn among lowland Maya groups, but it is possible that the timing followed in Chichipate is quite generalized. Reina discovered in San José, for example, that late afternoon is usually chosen as the best time to commence firing, for then winds are likely to be calm (Reina, 1967: 4).

Their basic task was to collect the clearing fee, see that the fire did not spread into adjacent land, and, where necessary, build firebreaks. When *comisionados* see that the fire is indeed spreading, they descend upon it with long, green branches and beat at it furiously.

Firebreaks and constant vigilance make uncontrolled fires rare. On occasions when fire does get out of control, a general call for help goes out, and not only the *comisionados* but everyone available gathers to rapidly open up still more firebreaks. This activity, combined with beating by sticks, usually controls the situation within the day, though large tree trunks, stumps, and roots may continue to smoulder for a week or more.

As with the Tajín Totonac (Kelly and Palerm, 1950:110), dried palm fronds or corn stalks are the most common forms of tinder. The fire is set *(xloch bal li xam)* simultaneously by all men involved in firing a particular field. For milpas set on hills, the fire is usually started at the top, so that it will gradually advance toward the bottom. Level milpas are generally fired in the direction of the wind *(yō chi chālc li ikʾ)*. Only men engage in the firing process. The whole attitude toward burning is rather casual; no ritual is performed. Stumps are left wherever they appear; no special attempts are made to remove them. Yet most milperos have rather successful burns.

Refiring

In those cases where fields have not been cleared well with the first firing, milperos try their skill one more time. This second firing *(xcab cʾatoc)* can occur at any time after the heat from the first fire has disappeared. It is limited to burning whatever debris has not yet been consumed. Seldom is any effort made to gather debris together for this second burning, however. Nor is there usually any third attempt at burning.

Some milperos wait until the day they are dibbling a plot to burn whatever debris remains. If it does not burn at this time, they simply leave that portion of the milpa unplanted.

Little opprobrium is associated with an unsuccessful burn. One milpero slashed a field late in the season, and proceeded to fire it while it was still green. It burned so poorly that he simply abandoned it and curtailed his planting activity by that much. He was not ridiculed. The general attitude prevails that "if it burns, it burns; if not, you cannot do anything about it."

Hardy Survivals

As was mentioned in the section on protective clearing, some plants are highly resistant to fire, and tend to revive once the rains have begun. Among domestic species, these include *ox* (*Alocasia macrorhiza* Schott), *tzʔin* (*Manihot esculenta* Crantz), *utzʔajl* (*Saccharum officinarum* L.), *tūl* (*Musa paradisiaca* L.), *quenēy* (*Musa sapientum* Kuntz), *ic* (*Capsicum frutescens* L.), and *is* ((*Ipomoea batatas* (L.) Lam.)). Of the wild plants, the *yaxteʔ* (*Brunfelsia* sp.), *akʔl* (Moraceae), *cakaj* ((*Bursera simaruba* (L.) Sarg.)), *pōk* (*Spondias purpurea* L.), *mocōch* ((*Orbignya cohune* (Mart.) Dahlgren ex Standl.)), and *saksi* (*Annona* sp.?) easily survive. Within six weeks after firing, the stumps and remains of these wild plants will sprout anew, and weeding during the first season after cutting over mature forest is limited to lopping off these *tux,* or sprouts.

Labor Requirements

The time required to clear firebreaks, set the blaze, and control it once it gets started, seems to be a function of five basic variables: the type of vegetative cover, the extent to which the debris has dried, the lay of the land, the nature of surrounding vegetation, and the number of men cooperating in the firing process. The average number of man-hours spent in firing, clearing firebreaks, and controlling the blaze was twenty-seven per *manzana* in 1965. In those cases where many men cooperated in the burning process, the time required to burn a *manzana* of land was reduced, though the total man-hours expended in the effort increased. Where six to ten men worked together, firebreaks could be constructed, and a *manzana* could be completely fired in an average of three hours and forty-five minutes. Where from one to five supervised burning, the average time required was eleven and one-half hours. From the standpoint of man-hours, smaller work groups appear to be more efficient than larger ones. When from one to five men controlled the burning process, the average man-hours required for 1 *manzana* of land was twenty-one hours and forty-five minutes, when from six to ten men were involved, the total required man-hours rose to thirty-eight hours and twenty-five minutes. But in all cases, the time requirements for building firebreaks and burning over debris are minor when compared to those of other activities in the milpa cycle.

Cropping

Once undesired vegetation has been eliminated through the activities of slashing, felling, leveling, and burning, the stage is set for beginning the cropping phase of the milpa cycle. Though this is dominated by maize, other plants are intercropped in such a fashion that a given milpa may continue to produce foodstuffs up to three or four years after being cleared. The four basic stages of milpa cropping are:

1. *C?at c?al*—Wet season milpa (literally "fire milpa"). Essentially sown to maize.
2. *Al ox, ch?op, tz?in,* etc.—New intercropped malanga, pineapple, manioc, etc. (*al* = "tender").
3. *Chek ox, ch?op, tz?in,* etc.—Old intercropped malanga, pineapple, manioc, etc. (*chek* = "old").
4. *Sak?icuaj*—Dry season milpa. Sown basically to maize (literally "summer milpa").

Intercropped items are managed in such a way as to require attention only during slack periods in the maize cycle. Table 9, which designates various crops, their planting, cultivation, and harvest periods, will help clarify the way in which the time sequence of these crops is coordinated. In addition to the five varieties of maize cultivated in Chichipate in 1965, twenty-one basic plant types were intercropped, and six were planted exclusively in house yards.[23]

In presenting this material, the form developed by Conklin in his *Hanunóo Agriculture* will be used, adding only minor variations. This permits a direct comparison of the two systems, clarifying immediately how much less complex is that of Chichipate. Information on each domestic plant is broken down into three basic categories: identification (columns 1-3), usage (columns 4-5), and planting data (columns 6-11).

Crop Identification

Kekchi basic plant name.—All entries are arranged alphabetically according to Kekchi basic crop names. The categories so designated often include two or more mutually exclusive subcategories or specific plant types, the number of which is noted in parentheses after the basic name.

23. Perhaps because of Chichipateños' pioneering status and their concomitant lack of familiarity with the lowland tropical forest environment, the variety of cultivated plant material they commonly identify is patently smaller

Botanical name.—The Latin binomials given in the third column are based on dried plant specimens collected in Chichipate in 1965. Several of the subcategories recognized by the Kekchi have not been identifiable on the basis of dried plant material, and there is a question as to whether our classificatory system would admit them, even if more adequate material were available.

English name.—English names most often associated with the designated plant segregates in the literature.

Crop Usage

Following Conklin (1957), several abbreviations have been employed to designate the use to which various plant forms may be put.

Food.—The specific manner in which the item is prepared and consumed has been indicated as follows: M, main starch staple; S, side dish, including spices, flavoring, and food coloring; s, snack food, i.e., that which is commonly eaten uncooked; N, no food value.

Other uses.—Other human needs met by the plant type: M, medicinal; R, ritual; T, technological; S, social; E, economic; C, cosmetic; N, none observed or recorded. As with the report on the Hanunóo, "T" refers to crops which provide materials needed in firemaking, cooking, and in the construction of tools, containers, house and agricultural equipment, hunting and fishing equipment, clothing, structures, musical instruments, and other useful objects. "S" refers to plants which furnish materials used in many social activities, such as fiestas and family visits. "E" is used to designate those items used commonly in trade. "C" designates those items used for personal beautification.

Planting Data

Incidence.—This category refers to the frequency and extent to which particular plant types are grown in new milpas: U, universal, i.e., planted widely in every milpa; W, widespread, i.e., planted in more than 35 per cent of the new milpas; O, occasional, i.e., planted in fewer than 35 per cent of the new milpas.

---

than that identified and used by other groups of Middle America that have long inhabited the lowlands. The reader is referred to Wisdom's book on the Chorti, as well as to Kelly and Palerm's work on the Tajín Totonac (Wisdom, 1940; Kelly and Palerm, 1950). In the latter book, an herbarium catalog of 367 plant specimens is provided.

Planting time.—The abbreviations refer to the preferred planting months. Months cited one-half year or more apart indicate that normally there is more than one annual crop. The letter "C" indicates that factors other than the calendar date are the principal determinants in selecting a time for planting.

Planting locations.—The preferred terrain, within the milpa, for planting: M, near the inner margin of the milpa; O, in the open, or central part of the milpa; S, at the base of stumps left in the milpa; D, in depressions, whether natural or left by rotted stumps; G, *guamil,* i.e., in low, first season fallow; H, house yard (outside the milpa).

Planting technique.—D, with a dibble stick *(auleb)*; P, with a planting stick *(macān)*; M, with a machete *(chʔīchʔ)*; F, with the finger *(ruʔuj ukʔ)*; S, by sowing or broadcasting *(xpajīnc iyaj)*. Each of these techniques differs considerably from the others. With the first, dibbling, the object is to punch, but not enlarge, a shallow hole into which seeds are dropped. With the second, using a planting stick which is usually broad and metal tipped, the worker turns over a circle of earth from 2 to 4 feet in diameter. He must dig rather deeply, for such things as plantain cuttings are large and must be firmly secured in the ground. In the third technique the worker uses a machete and tries to stir well a shallow piece of ground into which small cuttings may be placed. The fourth technique, finger planting, is limited to marginal crops, such as peanuts and garlic, and is employed only in soil which has become soft and friable through recent slashing and burning. Broadcast sowing, the final technique, requires the same type of soil preparation as does finger planting and is also limited to marginal crops. A few house yard plants such as carrots, radishes, and cabbage are often thus sown in well-prepared seed beds. These are kept shaded with palm leaves, and irrigated daily; otherwise they fail to germinate.

Plant spacing.—This is designated in three possible ways: (1) the actual average minimum spacing that has been ascertained through interviews and measuring, (2) planting in patches for which there is no standard spacing, and (3) isolated planting of a single or very few specimens:

O.O., actual average minimum spacing
P, planting in patches
S, isolated specimens.

Linear units of measure employed in calculating the space between plants are only approximate, for they are based on body pro-

TABLE 9
PLANTED MILPA CROPS

Food	*Other Uses*	*Incidence*	*Locations*
M —Main starch staple S —Side dish s —Snack food N —No food value	M —Medicinal R —Ritual T —Technological S —Social E —Economic C —Cosmetic N —None observed	U —Universal W —Widespread (more than 35 per cent) O —Occasional (less than 35 per cent) *Technique* D —Dibble P —Planting stick M —Machete F —Finger S —Sow (broadcast)	M —Margin O —In open, central part of milpa S —At base of stumps D —In depressions G —In first season fallow (*guamil*) H —House yard *Spacing* O.O—Actual minimum spacing P —Planting in patches S —Isolated specimens

Kekchi basic plant name	Botanical binomial	English name	Uses		Planting Data					
			Food	Other	Inci-dence	Time	Loca-tion	Tech-nique	Spac-ing	Propagation method
Ānx (2)	*Allium sativum* L.	Garlic	S	MRSEC	O	June July Aug.	O	F	5″	Peel
Arroz (2)	*Oryza sativa* L.	Rice	MS	SE	O	Apr. May	O	D	21″	Seed
Carabanzu (2)	*Vigna unguiculata* (L.) Walp.	Chick pea	Ss	RSE	O	May	MS	D	P	Seed
Cebolla (3)	*Allium ascalonicum* L.	Onion	S	MRSE	O	May June July Aug.	O	D	5″	Bulb

Che? quẽnk? (1)	*Cajanus bicolor* DC.	Tree bean (Pigeon pea)	S	RTSE	O	May June	H	D	P	Seed
Ch?op (2)	*Ananas comosus* (L.) Merrill	Pineapple	Ss	MSE	W	June July Aug.	D	M	15″	Slip in rows
C?um (3)	*Cucurbita* sp.	Squash	S	MSE	U	May Oct. Nov.	O	D	P	Seed
Ic (5)	*Capsicum frutescens* L.	Chili pepper	S	MRSEC	U	Nov	MG	D	21″	Seed
Is (3)	*Ipomoea batatas* (L.) Lam.	Sweet potato	Ss	SE	U	C	O	M	P	Slip in rows
Ixim (5)	*Zea mays* L.	Maize	M	RTES	U	May Oct. Nov.	O	D	35″	Seed
Lamunx (1)	*Citrus aurantifolia* (Christm.) Swingle	Lemon	s	SE	O	C	H	F	S	Seed
Mani (1)	*Arachis hypogaea* L.	Peanut	s	SE	O	June July	OM	F	6″	Seed
Marañon (1)	*Anacardium occidentale* L.	Cashew	s	MSE	O	June	G	D	S	Seed
May (2)	*Nicotiana tabacum* L.	Tobacco	N	MSE	W	Aug.	M	S	39″	Seed; transplant to 39″ in November

TABLE 9—*Continued*
PLANTED MILPA CROPS

Kekchi basic plant name	Botanical binomial	English name	Uses		Planting Data					
			Food	Other	Incidence	Time	Location	Technique	Spacing	Propagation method
Ox (1)	*Alocasia macrorhiza* (L.) Schott	Malanga	MS	SE	U	C	M	P	38″	Tuber
Papaya (1)	*Carica papaya* L.	Papaya	s	SE	O	June and later	M	S	P	Seed
Pata (1)	*Psidium* sp.	Guava	s	MSE	O	June	H	D	S	Seed
Pepinu (1)	*Cucumis sativus* L.	Cucumber	s	SE	O	May June July	M	D	37″	Seed
Pix (3)	*Lycopersicon pimpinellifolium* Mill.	Tomato	Ss	SE	O	Feb. July Aug.	G	S	P	Squeeze seeds from ripe fruit directly on the ground.
*Quēnk*ʔ (3)	*Phaseolus* sp.	Bean	MSs	RSE	W	May	OSM	D	P	Seed
Queneȳ (3)	*Musa sapientum* Kuntz	Banana	s	SE	O	C	O	P	111″	Sucker in rows
Rabano (1)	*Raphanus* sp.	Radish	S	SER	O	June	H	Planted in rows in well-prepared seed bed. Kept shaded with palm leaf and irrigated daily.		

Repollo (1)	*Brassica* sp.	Cabbage	S	SER	O	June	H	S	18″	Seed; transplant to 18″ in July.
Sandia (3)	*Citrullus vulgaris* Schrad.	Watermelon	s	SE	O	May June July	M	D	37″	Seed
Tul (3)	*Musa paradisiaca* L.	Plantain	Ss	SE	O	C	O	P	111″	Sucker in rows
Tzʾin (2)	*Manihot esculenta* Crantz	Manioc	MS	SE	U	June Oct. Aug. preferred	D	M	92″	Hardwood cutting
Utzʾajl (3)	*Saccharum officinarum* L.	Sugar-cane	Ss	MRTSE	O	C	OM	M	68″	Slip
Zanahoria (1)	*Daucus* sp.	Carrot	S	SER	O	June	H	Planted in rows in well-prepared seed bed. Kept shaded with palm leaf and irrigated daily.		

portions and their multiples (Table 10). Ropes are held in the hands and measured off against fingers, elbows, breastbones, and shirt seams. Once measured, they are cut and knotted, thus becoming permanent units for marking off planting and boundary distances.

TABLE 10
KEKCHI LINEAL MEASURES

Unit	Average length (inches)	General Guatemalan equivalents	Basis for calculation (adult male)
Jun c'utub vār	8¼ (20.96 cm)	One-fourth *vara*	From tip of the fully extended fifth finger to the tip of the thumb.
Mero vār	17½ (44.45 cm)	One-half *vara*	From tip of extended middle finger to the elbow, with the arm fully extended.
Jun vār	35 (88.9 cm)	One *vara**	From tip of the middle finger to the breastbone, arm fully extended.
Jun mokoj	68 (172.72 cm)	One *brazada*	From tip of extended middle finger to tip of extended middle finger of the other hand, arms straight out to the right and left.
Jun metro	41 (104.14 cm)	One meter	With left arm fully extended, from the tip of the middle finger to the seam of the shirt, at the juncture with the right sleeve.

*Redfield found that in Yucatán, as in Chichipate, the *vara* was defined as the distance from the fingertip to the middle of the breastbone (Redfield and Rojas, 1962: 43).

Summary of Tabulation

Of the 60 specific and 28 basic cultivates commonly found in Chichipate, all but six are planted in milpas. This leaves about 25 per cent of the basic cultivates and only 10 per cent of the specific cultivates for house yard plots. Starches represent the most common plant type grown in milpas and serve as the dietary base. Maize far outstrips all other plant species in importance; though only five varieties are grown in Chichipate, the maximum for varieties of other plant species never surpasses three.

Six crops are grown exclusively in house yards: pigeon pea *(che? quēnk?)*, lime *(coc? lamunx)*, carrot *(zanahoria)*, radish *(rabano)*, cabbage *(repollo)*, and guava *(pata)*. All others may be found both in milpas and house yards; none is limited to milpa cultivation alone.

With the exception of maize, there is not a single plant which is grown in every milpa, yet of the twenty-two basic cultivates grown there, fourteen clearly predominate over others in popularity and frequency.

STARCHES	LEGUMES
Grains	Beans *(quēnk?)* (3)
Maize *(ixim)* (5)	Chick peas *(carabanzu)* (2)
Tubers	VEGETABLES AND FRUITS
Manioc *(tz?in)* (2)	Tomato *(pix)* (3)
Malanga *(ox)* (1)	Pineapple *(ch?op)* (2)
Sweet potato *(is)* (3)	Squash *(c?um)* (3)
Tree Crops	SNACKS, SPICES, AND CONFECTIONS
Plantain *(tul)* (3)	Sugarcane *(utz?ajl)* (3)
Banana *(quenēy)* (3)	Chili pepper *(ic)* (5)
	Tobacco *(may)* (2)

With the exceptions of maize, squash, and beans, all other crops are planted only in small patches. Though beans, chick peas, and squash are planted simultaneously with maize, it is seldom that even they are intended to "fill out" the milpa. Maize remains not only the basic dietary staple, but the predominant form of plant life in every first season field.

One of the adjustments migrants to Chichipate have had to make has stemmed from the difficulties that accompany bean production in the lowland tropics. Insect pests are so ubiquitous that the plants tend to be decimated long before harvest time. In 1965 milperos were still planting beans in the hope that, through sheer luck, a decent crop would ensue. Nor had they yet changed their dietary habits. Beans were served with nearly every meal and, to purchase them, the milperos were forced to lay out cash earned from the sale of maize.

After maize, the cultivates most commonly distributed throughout the milpas are beans, squash, chick peas, sweet potatoes, chili peppers, manioc, and malanga, in that order. Everything else tends to occur in only isolated patches, though two milperos had sown entire fields to sugarcane. Rice, not included in the list of commonly grown cultivates, is of relatively recent introduction. The idea of growing rice in the low-lying ground surrounding Chichipate appears to have

TABLE 11
MILPA CROP HARVESTING PERIODS

Crop	Jan.	Feb.	Mar.	Apr.	May	June	July	Aug.	Sept.	Oct.	Nov.	Dec.
Banana	x	x				x	x	x	x	x	x	x
Bush bean									x			
Cabbage			x	x	x			x	x	x		
Carrot			x	x	x			x	x	x		
Chick pea									x			
Chili pepper	x						x	x	x	x	x	x
Garlic						x	x					
Lime										x	x	x
Maize		x	x	x				x	x	x	x	
Malanga	x	x	x	x	x	x	x	x	x	x	x	x
Manioc							x	x	x	x	x	x
Onion											x	x
Papaya								x	x			
Pineapple			x	x	x	x	x	x	x			
Plantain	x	x				x	x	x	x	x	x	x
Pole bean	x	x	x	x	x							x
Radish			x	x	x			x	x	x		
Rice									x	x	x	
Squash and pumpkin	x								x	x	x	x
Sugarcane	x	x	x	x	x	x	x	x	x	x	x	x
Sweet potato	x	x	x	x		x	x	x	x	x	x	x
Tobacco	x	x	x									
Tomato	x	x	x	x	x	x	x		x	x	x	x
Watermelon				x	x	x	x	x	x	x	x	x

come from experience milperos have gained working in the rice fields of Yuscarán. In 1965 two milperos planted entire milpas to this cultivate as an experiment. A few others dibbled rice seed into water-logged depressions in the middle of maize milpas. There was considerable speculation as to whether the cultivate would become a major crop for the community in the future. It was still a very rare dietary item in 1967, however.

Maize, the major milpa crop, is planted twice a year. Simultaneously, beans, chick peas, and squash seeds are dibbled into the ground. Other cultivates which are sown under the maize, once it has begun to grow, are tomatoes, watermelons, and chili peppers. A number of common cultivates may be sown at any time throughout the year, the conditions depending more on the soil, vegetative cover, and lay of the land than on the factor of time. These include manioc, plantain, banana, sweet potato, and malanga. Planting periods for the remaining crops tend to fit nicely between the times of heightened activity in the seeding and harvesting (Table 11) of major cultivates. Although there are certainly lulls of milpa activity, there is hardly a month when some productive work is not done.

No milpa contains all of the basic cultivates, and certainly none approaches having all the specific ones. The maximum number of basic cultivates found in any given milpa is eighteen. In the rainy season, except for the rare cases when an entire milpa has been planted to sugarcane or rice, maize is the predominant cultivate, and it shades most others. In the dry season, however, entire fields of sweet potatoes, beans, and chili peppers, sown independently from maize, may occasionally be seen. Otherwise, cultivates are intercropped either in milpas of maize or are scattered throughout swiftly replenishing secondary growth.

Cʔat cʔal *Making*

Maize planting ritual.—An elaborate series of ritual taboos and practices emphasize the primary importance of maize in the Kekchi economy. While not every milpero observes them all, most try to, and the first man in the village to begin planting is *required* to do so. The observances begin nine days before seeding, from which time sexual intercourse is forbidden. The first night of these rituals, prayers are said both in the home and at the *ermita*. One of these, recorded verbatim, gives considerable insight into the supernatural dimension of maize farming, as perceived by Chichipateños:

Prayer of Preparation

Kekchi	English
Ay Dios Jesucristo, anakcuan xinchal rubel ācuok rubel ācuukʔ. Ma us ma incʔaʔ tinye a cuaʔ. Ay Dios, Padre Eterno, nequeʔxye li Kanaʔ kayucuaʔ xeʔquirisin ke, nequeʔxye nak lokʔ li kacua li kucʔa. Ut anakcuan xinchal xpatzʔbal, xtzʔamanquil rubel lā cuok rubel lā cuukʔ, chi anchal inchʔōl li cuām. Anakcuan tinpatzʔ li santil	Oh, God, Jesus Christ, now I have come under your feet, under your hands. It is good or not (what) I will say, Father. Oh God, eternal Father, they say, our mothers and fathers that reared us, they say that our food and drink are precious. And now I have come to ask, to beseech under your feet, under your hands, with all

maʔaj chicuix chincʔatc, li santil maʔaj aj Jehova, santil maʔaj aj San Rafael, santil maʔaj aj San Miguel, oxib chi naʔaj ninpatzʔ chicuix chincʔatc. Cʔaʔru aj e cʔaʔru xcuanquil nak tinpatzʔ aʔan chirix lin cua li cuucʔaj haʔ xnumsinquil li cutan sakʔēnc. Joʔcaʔin anakcuan saʔ li beleb cutan anakcuan nak tināuk. Entons tinye ācue anakcuan tacuy inmāc tana, bar tana us li cuan tinye ācue anakcuan at incuaʔ. Tinpatzʔ ajcuiʔ lā santil apōstl, cablaju chi apōstl xbec chacuix nak xatyoʔla Jerusalem. At innaʔ at inyucuaʔ ay Dios a cuaʔ, anakcuan tintzʔama saʔ ācuok saʔ ācuukʔ chijunil lā bendicion, tācuānk lā musikʔ. Nacuabi resil lā cuām lā musikʔ, li Dios Espiritu Santo. Oxib vuēlt tinye nak saʔ xcʔabaʔ li Dios Acuaʔbej, li Dios Cʔajolbej ut li Dios Espiritu Santo, amen. Oxib vuēlt, saʔ xcʔabaʔ li Dios Acuaʔbej, li Dios Cʔajolbej, li Dios Espiritu Santo, amen. Saʔ xcʔabaʔ li Dios Acuaʔbej, li Dios Cʔajolbej, li Dios Espiritu Santo, amen. Ay Dios a cuaʔ xtzʔakloc lā bendicion, lā cuām lā musikʔ bar cuan xinye. Anakcuan bar xtʔanec re lin cua li cuucʔa haʔ? Chire chiru li lokʔbelaj tzūl, lokʔbelaj takʔa. Yoʔyoqueb aʔan. Nequeʔxye nak cuan nayehoc re nak incʔaʔ moco yoʔyo ta. Ay, incʔaʔ. Caʔajcuiʔ li cuan saʔ choxa, chanqueb. Incʔaʔ. Lāo, tinpatzʔ ajcuiʔ li lokʔlaj tzūl, lokʔlaj takʔa, jarub li ninnau xcʔaba, aʔan ajcuiʔ tinpatzʔ. Anakcuan bar cuan li tzʔakl xakxo saʔ xben li kacua li kucʔaj haʔ. Caʔajcuiʔ li kacuaʔ San Pablo, kacuaʔ Cojaj, li kanaʔ Itzʔam, kacuaʔ Cʔancʔamas, kacuaʔ Chixbajau, kacuaʔ Paxucuac, kacuaʔ Xacʔoj, kacuaʔ Cuaxpom, kacuaʔ Jolom Hix, kacuaʔ Secʔāche, kacuaʔ Sebacaron, kacuaʔ Isabal, kacuaʔ Sebacari, kacuaʔ Santasabel. Caʔajcuiʔ aʔin tzʔakal xpuērs li cuātin a cuaʔ. Nak

my heart, my soul. Now I ask the holy place behind beside me, the holy place Jehovah, the holy place San Rafael, the holy place San Miguel, three places I ask behind beside me. For what purpose, for what power that I ask this concerning my food and drink for passing the days. Like this now on the ninth day I will plant. Now I say to you perhaps you will forgive my sin. Perhaps it is good what I will say to you. My Father, I also ask the holy apostles, twelve apostles walked behind you when you were born in Jerusalem. My mother, my father, oh God Father, now I beseech in your feet, in your hands all your blessing. May there be your spirit. I hear the sign of your heart, your spirit, God Holy Spirit. Three times I will say in the name of God the Father, God the Son, and God the Holy Spirit, amen. Three times, in the name of God the Father, God the Son, God the Holy Spirit, amen. In the name of God the Father, God the Son, God the Holy Spirit, amen. Oh, God, Father, it is completed your blessing, your soul, your spirit which I said. Where are my food and drink falling? Beside, in front of the holy mountain, the holy below. They are alive. They say that there are those who say that they are not alive. Oh, no, only the one in heaven they say. No. We, I ask also the holy mountain, the holy below, all whose names I know, those also I ask. Where is the one who really controls (sufficiently stands on) our food and drink. Only our father Saint Paul, our father Cojaj, our mother Itzʔam, our father Cʔancʔamas, our father Chixbajau, our father Paxucuac, our father Xacʔoj, our father Cuaxpom, our father Jolom Hix, our father Secʔache, our father Sebacaron, our father Isabal, our father Sebacari, our fa-

ninnau ta cuan xjunil li bar cuan xcanabeb xcʾabaʾ chak li lokʾlaj tzūl, li lokʾlaj takʾa, tinye ajcuiʾ joʾ ta bar cuan li nacutun chicuu. Caʾajcuiʾ aʾan tinye anakcuan a cuaʾ. Joʾ aʾin saʾ beleb cutan nak tinãuk. Ay Dios a cuaʾ caʾajcuiʾ aran xorakeʾ. Ay Dios tinpatzʾ xcʾabaʾ anakcuan tinpatzʾ xcʾabaʾ li apōstl San Juan, San Pedro, San Pablo, San Lucas, San Marcos, San José, San Vicente, San Antonio, Salvador del Mundo, Salvador Mio, San Antonio del mōnt, aʾan tzʾakal lix diosil li kacua li kucʾa. Saʾ xcʾabaʾ li Dios Acuaʾbej, li Dios Cʾajolbej, li Dios Espiritu Santo, amen. Saʾ xcʾabaʾ li Dios Acuaʾbej, li Dios Cʾajolbej, li Dios Espiritu Santo, amen. Saʾ xcʾabaʾ li Dios Acuaʾbej, li Dios Cʾajolbej, li Dios Espiritu Santo, amen.

ther Santasabel. Only that is the sufficient strength of my words. If I knew all the names that the Holy mountain, the holy below, left, I would say them also as I said the ones that appear to me. Only this I will say now, father. Like this in nine days I will plant. Oh, God, Father, only there we finished. Oh God, I ask (in) their names now. I ask (in) the name of the apostles, San Juan, San Pedro, San Pablo, San Lucas, San Marcos, San José, San Vicente, San Antonio, Salvador del Mundo, Salvador mio, San Antonio del monte. These are the real gods of our food and drink. In the name of God the Father, God the Son, and God the Holy Spirit, amen. In the name of God the Father, God the Son, and God the Holy Spirit, amen. In the name of God the Father, God the Son, and God the Holy Spirit, amen.

On the third night before seeding, men and their wives again walk to the *ermita,* burn candles and copal, and pray for safety, a good crop, and a long and good life. This visit may occur either at dusk or the following dawn. While the man prays, the woman, who has brought the brazier full of smouldering copal, shakes it as an offering to the saints and to the milpa.

As the two enter the picketed "sanctuary" area, each makes the sign of the cross. Then the woman gives her husband candles, which he places before each of the saints. He directs prayers to the saints, to God, and to the hills, that his plants grow strong and healthy, and that there be no accident or illness during the growing season. He and his wife continue burning candles and copal until up to three large or nine small candles and a quarter pound of copal have been consumed. More candles and incense, it is believed, would create so much heat that the milpa would turn yellow before it was ripe.

From one to two hours after having gone to the chapel, the two return home, and there they burn candles and copal before the saint of the house where they have already set the bag of seed. The cross placed in the milpa at planting time has hopefully survived the fire (the ground around it had been carefully cleared), and is now brought and placed on top of the open sack of maize. Care is exercized that no one except the milpero and his wife touch the seed, for to do so would be to betray the reverence it is due. The same prayers that were used in the chapel are here again offered.

The sacredness of the occasion may be appreciated through one verbatim, summary account of this phase of the ritual:

> In three days I will plant. Then I will go to the holy church. I will go to talk to our Father. He is in the world. Is it he that I will not find the name of the one which I asked? I exalted him completely. Then a hundred thousand times he forgave, he lost my sin. He, Jesus Christ and Jehovah and San Rafael, San Miguel. And I ask again (in) their names the apostles San Juan, San Pedro, San Lucas, San Marcos, San José, San Vicente, San Antonio, Saviour of the world, my Saviour, and San Antonio of the mountain. I ask in their names a second time when (in) three days I will plant.

For three successive nights this ritual is repeated. During this time, adults of the family abstain from manioc, sweet potatoes, bread, and tree-borne fruits. The belief reinforcing this practice is that if one eats any of these foods, the maize will fill with worms, for all these foods are worm-prone. Before maize may be planted, an all-night wake *(yoʔlec)* must be held. Some families hold it on three successive ritual nights. Others limit it to only the third.

The wake begins at about 7:00 P.M. Again candles and copal are offered to the household saint. Women also grind green *cacao* (chocolate), mix it with water, place it in a gourd or cup, and set it on top of the maize seed. They add to this offering chicken broth made with pieces of chicken meat, garlic, salt, and chili pepper. On top of the bowl of broth, they place a stack of tortillas. The whole is set on the open sack of maize seed. Some families complete their offerings by killing a turkey or, preferably, a pig, and sprinkling the blood over the seed. As they do so, they pray to the surrounding hills and to God to give them permission to enter into the milpa, to grant them a successful harvest, to protect the milpa against marauding animals, and to liberate them from the threat of bad airs *(aire)*. These requests they make in the name of the Holy Spirit.

Once the spirits and maize have been fed, bitter chocolate *(cacao)*, chicken broth *(cālt)*, tortillas *(cua)*, and sometimes tamales *(oben)* and *atole (ukʔun)* are served until everyone present has eaten his fill. The food offered to the saints and to the seed is, at the end of the ceremony, given to children. It is believed that their innocence or lack of sin makes them the only persons who can eat it without harm.

Around six or seven in the morning, the work party begins to gather. They are first given food, and then the seed is divided among them. The entire group proceeds to the milpa; the "owner" of the plot arrives first. He goes immediately to the center of the field, kneels, and replaces the stick cross that had been resting on top of the sack of seed. In front of this cross he places a lighted candle, burns more copal, and prays. When the candle has burned down, he plants the first seed in its place, and this spot becomes the axis for a cross which will be composed of six hillocks of maize. The order in which this cross is sown is (1) axis, (2) left transept, (3) right transept, (4) head, and (5) and (6) foot. Into each of these hillocks should go exactly six grains.

	.4	
.2	.1	.3
	.5	
	.6	

While this ceremony is being executed, the other members of the work party are off in the bush looking for or sharpening their dibbles. The "owner" of the plot has looked for and sharpened his some days before. Planting activity begins only after the ceremony has been completed and all have secured their dibbles. It continues until around 11:00 A.M., at which time the workers begin to walk back to the house of the milpa "owner." There, in typical festive work-party fashion, they are feted with a banquet of tortillas, roast pig, tamales, and bitter chocolate. Before the food is distributed, copal is burned and prayers are offered that no one fall ill from the meal. Of the foods, chocolate is the most esteemed, and four or five gourds-full are placed on the table before the saint. Though it is from these that the work party drinks, one gourd is left untouched, as an offering for the saint. As all foods are brought from the fire, they are placed first in front of the saint and consecrated with copal fumes before they may be distributed to the guests.

Except for a taboo on sexual intercourse that continues for nine nights afterwards, this festive meal is the final ritual act in planting maize. Following it, the work party returns to the fields, the ideal being to complete the seeding of a given milpa in a single day. Aside from the food and sex taboos already mentioned, there appear to be no other important ones connected with maize planting. The phase of the moon holds no importance for the act. Care is taken, however, not to hang a bagful of maize seed from a tree, for this is believed to bring an infestation of wasps. Seed bags, when they are taken from the shoulders, should be set on the ground.

Preparation of seed.—Because milperos try to plant maize within one week after burning, and because seed should be ready at least a day or two before the planting wake, seed selection begins as soon as the firing has been completed. Ideally, rains will begin just after planting. Several weeks of dry weather can follow seed time, however, without causing undue concern on the part of the milperos. During such a dry spell, they conceive of the seed as merely resting in the ground and suffering little damage. Reina, on the basis of his San José data, speaks of the sweating of the soil in relation to planting. According to him, seeding can take place as long as eight days after burning because the "sweat" is effective for that period of time (Reina, 1967: 5). No such concept was found in Chichipate.

The five types of maize grown in Chichipate are presented in Table 12. They correlate remarkably well with maize varieties grown elsewhere in the Maya area. In general, Maya farmers stick to a very few varieties, though they may be aware of more (Hester, 1954: 161), and these are customarily distinguished by color. The basic ones recorded for Yucatán by Hester are white *(zac)*, black *(ek)*, yellow *(kan)*, and red *(chac)*, classifications almost identical to those used in Chichipate today. As Hester points out, it seems more than coincidental that these four colors are those most prominently associated in the Maya Pantheon, and that they are also associated with the cardinal directions (Hester, 1954: 161).

Seed selection is done by the milpero and his wife. They place each separate variety of maize into a different gunny sack, for they believe that mixing the seed tends to lower crop yield. Each grain must be inspected individually, and is selected for planting only if it has a "son." This "son" *(ralal ixim)* is the inner, lighter core of the grain, and must be intact if the seed is to sprout. For each *manzana* of milpa, about twenty-five pounds of seed are prepared, no matter what the lay of the land may be.[24] Kekchi measures used

24. Kelly and Palerm uncovered a complex method of seed preparation among the Tajín Totonac that seems to have no parallel in Chichipate. "Be-

in calculating planting needs result from a combination of Iberian with indigenous standards. Table 13, derived from informant statements and checked by standardized Western measures of weight and volume, should be considered only approximate, since Kekchi implements for measuring include such variable items as gourds, cups, basins, and bags.

Seed is divided among the members of the work group before they leave for the field. Each man carries it in his shoulder bag. Two *jōm* are enough for one man's half day of labor. At noon, when the worker returns for the festive meal, he refills his bag.

The bag of maize seed usually stays in the milpero's house. There the maize is mixed with bean *(quēnkʔ)*, chick pea *(carabanzu)*, squash *(cʔum)*, and/or calabash *(cʔum)*. About two pounds of bean seed, two of chick pea, and one-half of squash or calabash are used for an *arroba* of maize. These intercropped seeds are mixed by hand in the gunny sack.

Should smaller portions of seed be prepared, for every pound of maize a handful of each of the other seeds is mixed in. Often the milpero is negligent in mixing seed, and the members of the planting party themselves must do the mixing just before leaving for the field. In these cases, they simply pick up a handful of each of the other seeds and add this to the shoulder bag full of maize. Maize yields may surpass 100 to 1 on good, well-drained soil.[25] Since there is no strict control over the proportion of intercropped seeds added to the maize, the yield of beans, chick peas, squash, and calabash is highly unpredictable.

Planting equipment.—Equipment used in planting maize is exceedingly simple, being limited to shoulder bags for holding seed, and dibbles *(auleb)*. The shoulder bags are customarily bought

fore planting, the corn is shelled and is placed with water in a wooden tray. If the container is of harder material, such as pottery or metal, the resulting maize crop 'will be hard, and difficult to grind on the metate.' On top of the soaking corn are stuck two unlighted candles . . . each supported by a corncob.

"The following day, the seed corn is placed in a box lined with leaves of banana or of *papatla* . . . and more leaves are used as a cover. The box is moved to the sun for a day, and with the heat, the corn begins to sprout. When the root is visible, the maize is sprinkled with a solution of creolin or coal oil to protect it from ants . . ." (Kelly and Palerm, 1950: 110).

25. This estimate, obtained from informants, accords remarkably well with Kelly and Palerm's statement for the Tajín Totonac that "it seems likely that the yield per liter of seed generally is at least 100 to 1" (Kelly and Palerm, 1950: 120).

TABLE 12

COMMON MAIZE TYPES IN CHICHIPATE

Kekchi name	Botanical identification	Treatment	Preference	Use
Nīnk saki hal (Large white corn)	Resembles closely the race *Tuxpeño* of eastern Mexico. Probably Mexican in origin, and may have been brought in several years ago when local corn failed.	Most widely grown. Late harvest.	Most preferred.	Tortillas, *atole,* tamales, animals, sale.
*Saki coc*ʾ *hal* (Small white corn)	Nal-Tel Blanco, Tierra Baja mixed with Tepecintle.	Most widely grown of the early maize. Harvested before *nīnk saki hal.*	Second in order of preference because of early harvest date.	Tortillas, animals, sale.
Maiz brinco (No translation, possibly "gringo" corn)	Closely related to Tepecintle which probably had its origin in Guatemala and is the progenitor of the Mexican *Tuxpeño.* White.	Widely cultivated. Late harvest.	Third in order of preference.	Tortillas, *atole,* tamales, animals, sale.
*K*ʾ*ani coc*ʾ *hal* (Small yellow corn)	Nal-Tel Amarillo, Tierra Baja.	Eaten especially when tender. Early harvest.	Fourth in order of preference.	*Atole,* tamales, but not tortillas, for the mash is too soft. Eaten often as corn on the cob. No market demand.
*K*ʾ*eki hal* (Black corn)	Negro de Chimaltenango.	Planted in only small portions of the milpa. Early harvest.	Fifth in order of preference.	Tortillas, *atole,* tamales, animals, rarely sold. Used for ritual.

TABLE 13
KEKCHI MEASURES OF VOLUME AND WEIGHT
USED IN MILPA AGRICULTURE

Kekchi measure	Average	General Guatemalan equivalents	Other equivalents
		VOLUME	
Chʔīchʔ secʔ	400 cc	*pocillo*	chocolate cup
Jōm	715 cc	*huacal*	small gourd
Palangān	2500 cc	*palangana*	wash basin
Chʔina chacach	3600 cc	*canasto pequeño*	small basket
Cubēt	9550 cc	*cubeta*	pail
Nimla chacach	25,300 cc	*canasto grande*	large basket
		WEIGHT	
*Jun ukʔ**		*una mano*	Five unhusked ears of corn, five bananas, etc.
Jun ucʔal	25 lbs	*una arroba* (*una olla*)	25 pounds of shelled corn, or 40-60 ears of unshelled corn.
Nimla ucʔal	50 lbs	*dos arrobas* (*olla grande*)	50 pounds of shelled corn, or 100-150 ears of unshelled corn.
Quintal	100 lbs	*quintal*	100 pounds of shelled corn, or 200-300 ears of unshelled corn.
Jun okʔob	150 lbs	*una fanega*	1½ *quintales* of shelled corn, or 300-450 unshelled ears.
Cajón		*cajón*	Varying size, from one to twenty *arrobas*.

**Jun ukʔ* is really more a measure of volume than of weight. It is, however, conceived by the Kekchi to form part of the weight series.

either in stores or from traveling salesmen, for the women of Chichipate do little weaving. Dibbles may be made on the spot or saved from one planting episode to another.

Preferred wood for dibble making is that of the *acteʔ* plant ((*Astrocaryum cohune* (Wats.) Standl.)). Straight shoots some 5 feet tall and 2½ inches in diameter are needed for an adult's dibble. The shoot, harvested green, is peeled smooth, and then the point is sharpened. Preferred weight varies, of course, with the height and strength of the dibbler, each preferring the heaviest dibble that he can easily manage. This weight adds to the dibbler's thrust and makes penetration of the soil that much easier.

Only one type of point *(ruʔuj)* is prepared for maize planting, and it is of the conical type. These points wear down fairly rapidly. In ordinary, fairly friable soil, they must be sharpened once or twice a day; in stony milpas, they need sharpening much more frequently.

A variant of the maize dibble is used for planting smaller seeds such as those of chili peppers. This small dibble *(chʔina auleb)* averages only about 3½ feet in height. It has a blunted end which stirs only the very top of the soil layer.

Once planting has been completed, if any dibbles appear to be serviceable still, they are stored in the house rafters to be brought out for the next season. Well-cared-for dibbles have been known to last three years.

The digging stick, or *macān,* is approximately the same size as the large dibble, and in fact may be made from a discarded *auleb.* It differs principally in that the point consists of merely an oblique cut across the bottom, rather than in a rounded, pencil-type cone. Some *macān* are made, however, with a wide steel blade at the bottom, and the shafts of these are about twice as thick and heavy as are those of *auleb.*

Planting labor groups.—Cooperative work groups are larger for planting than for any other single activity in the milpa cycle. This is related partly to the need to get the maize into the ground quickly after burning, partly to the organizational advantages of dibbling an entire milpa in a single day, and partly to the festive air of expectancy that surrounds the planting process. Planting groups, in 1965, ranged from seven to thirty-five persons. Recruitment was as varied as the composition of the groups. All milperos invited their close consanguineal, ritual, and affinal kin. With them they simply exchanged labor for labor. But many also invited men who were no relation whatsoever. Some of these, personal friends, were drawn into planting activity on a reciprocal basis. Others were simply newcomers to the community who had time at their disposal, and who were willing to work for the 100 ears of corn a day that is customary pay.

Of ten milpa planting groups sampled in 1965, nearly half the participants (42 per cent) appear to have had no deeper commitment to the sponsoring milpero than that of friendship. The next largest category, 32 per cent, was made up of individuals who had migrated from the same Alta Verapaz community as had the milpero. Fifteen per cent were consanguineal relatives, and affinal relatives accounted for a surprisingly low 11 per cent, indicating perhaps that the ties imposed by post-marital bride service are of short duration in a place like Chichipate. The whole cooperative pattern points out how much milperos depend on unrelated friends and neighbors in the frontier setting. Whether this is a distinctly different pattern from that which exists in the Alta Verapaz communities from which these people have come is impossible to say until a study has been made of the latter. It would seem logical, however, that on the frontier, residence groups take over many of the functions of kin groups in a traditional, sedentary setting.

Dibbling and seeding.—Dibbling and seeding are carried out simultaneously and by the same person. The dibbler, if right-handed, grabs his instrument in his right hand, holds it in front of him with his arm in a slightly flexed position, and jabs the point about 5 inches into the ground. As he does this, he gives a slight movement to the left and right and to the back and front with the dibble, and withdraws it as straight and quickly as possible, so that the hole he has just punched will not collapse. Between holes he leaves spacing of approximately 35 inches.[26] Less space between hillocks is considered too close *(mas cocˀ xyānk)*, more as being too far apart *(mas najt xyānk)*. Both sandy *(samahiˀ ru li chˀochˀ)* and black *(kˀek li chˀochˀ)* soils are fairly easy to dibble, though quantities of charred sticks and poles may create difficulties in the latter. The most problematical soils are those that are either loose and dry *(chak li chˀochˀ)*, or wet and sticky *(sulul seb li chˀochˀ)*. With the former, the hole tends to refill before the seed can be introduced; with the latter, e.g., heavy, poorly drained clays,

26. Reina, for San José, says that good spacing ranges from a meter to a meter and one-half for each group of kernels (Reina, 1967: 6).

Dibbling and seeding.

both penetration and extraction of the dibble require more than the average expenditure of energy.

Between dibbling and seeding, a fast dibbler may complete about six hillocks a minute. Three is considered slow, and one unacceptable. When a milpero finds that one of his work party is slow, he takes mental note of the fact and, in his reciprocal return of labor, performs in like manner. If the slow worker is being paid a wage in kind or cash, however, the milpero has no recourse other than to refrain from hiring him again.

The planting party starts at the lower end or foot *(rok li cʔal)* of the milpa, to the right of the ritual cross. The "owner" works closest to the cross, starting just to its right, at the foot of the milpa. As

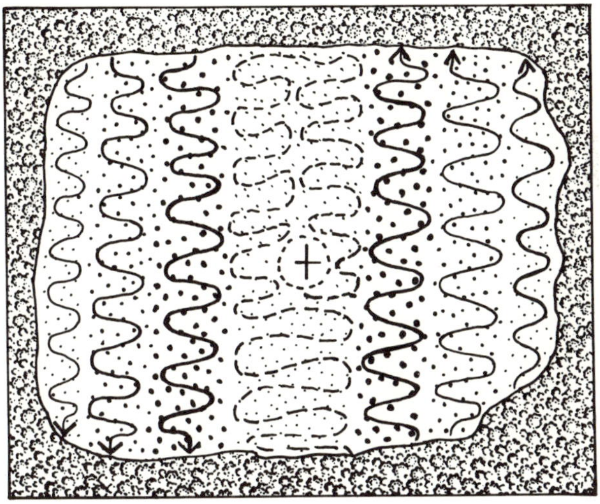

Dibbling pattern.

dibbling and seeding proceed, he works up the central section of the milpa to the ritual cross, circles around it, goes to the top of the milpa, and finally returns to a spot to the left of the base of the cross. In the meantime, the other dibblers *(aj acuinel)* plant the right side of the milpa, then switch to the left side, and finally finish at the left-hand foot of the field.

During their labor, most dibblers smoke homemade cigars, though there appears to be no obligation to do so. Competition exists, and slow workers are derided. More severe sanctions for sloppy and inefficient labor are limited merely to reluctance to cooperate on future occasions.

As soon as the dibbler has perforated the soil, he drops his seed into the hole. If the dibbler is right-handed, this seed is held in the left hand, and the fingers swiftly count off the number of grains before letting them fall. While they are being counted, the grains stay on the upper and middle phalanges of the fingers. They are dropped into the hole with a quick, downward snap. The ideal number of seeds for a given hillock of maize *(xtōnal)* is six (*cuakib ru li xtōnal*—"six stalks [faces] in the hillock"). However, a count of more than a thousand hillocks in various milpas revealed that the number of growing maize stalks in each ranged from two to eleven, the average being 5.99 in poor land and 5.01 in good land. Granted that some of the kernels may not have sprouted, nevertheless, it is clear that the ideal of six kernels for each hillock is often lost in the hustle of dibbling and seeding. The sample count indicates that the ideal pattern is found in only about one-third of the hillocks.[27]

Once the worker has dropped seed into the dibbled hole, he closes it with the dibble. He then proceeds to the right, left, or straight ahead, depending on his point of progress through the milpa. Only when he comes to stumps or water-filled depressions does he leave spots unseeded. He dibbles even the top of ant hills, though seldom do these spots yield much harvest even if the kernels sprout, another indication of the "inefficiency margin" in Chichipate agriculture.

Most dibblers continue to work until their second, or afternoon, bag of seed has been completely used up. This would appear to give an advantage to those individuals who carelessly drop more than six kernels in their hillocks. Usually, however, these are also the slow workers, and so they tend to finish at about the same time as the others. Once a worker has finished, he leaves alone and returns home.

A well-prepared milpa will be evenly dibbled and seeded. Boys are taught this ideal and are taken to planting parties from about the age of seven. Before going to the fields, they are carefully instructed in handling both the dibble and the seed. Keeping seed in the palm of the hand while dropping exactly six kernels onto the phalanges requires considerable skill, and boys must learn early to use the thumb in counting and in returning extra kernels to the palm.

27. The ideal of approximately six seeds in a single hillock seems to have both historic depth and geographical spread. For colonial Yucatán, Landa speaks of "five or six grains" (Landa, 1938: 111); for modern Yucatán, Morley writes of "three to six kernels" (Morley, 1947: 147); and for La Venta Island, Drucker and Heizer state that five kernels are normally dropped into each hole (Drucker and Heizer, 1960: 41).

By assigning workers to successive bands of a certain width around the perimeters of the milpa, an entire side is seeded at one sweep. Thus, when the last worker returns to the foot of the field, all dibbling and seeding have been completed. Replanting and transplanting are occasionally tried, but very few milperos do either on a regular basis. The general philosophy prevails that if the milpa did not germinate the first time, it will not a second, for something must be wrong with the site. "Why lose work two times?" say the milperos, and they simply abandon the field.

Neither is there any attempt to remove excess plants from a hillock of maize, nor to cut out the suckers that later appear. The results of planting are accepted in a fatalistic fashion rarely found in more exacting agricultural systems.[28]

This sense of fatalism seems closely related to milperos' perception of relevant forces in plant growth and development. The secular and sacred are intertwined so closely that, with a basic activity such as planting, it is very difficult to separate the two. A description given by one milpero of his planting party will illustrate:

> Then the day of my planting has arrived. I shelled my seed. Tomorrow is my planting so I shelled my seed. One day I hunted. A half day I hunted. I began to shell my seed. I burned my incense. I loosened my belt so that my ears of corn would not be hard. And then I went to do *costumbre* for my corn field. The planters are gathering. I went to do *costumbre* for my corn field. Then I came to do *costumbre* for my corn field, when the planters had already gathered. But I did *costumbre* for my corn field in the center of it. I took my companions. Then in the middle I divided. I put my companions to the right of me. I circled my planting. Where I began, there I finished. Then I arrived at noon. My wife had already burned incense. She had already asked God again for happiness for both of us. "Now have you come?" "We have come. Now let our dishes be gotten ready," we said. She began to get ready our dishes. Our dishes were ready. Then I burned

28. Chichipate planting procedures correspond closely to those described for the Tajín Totonac. Kelly and Palerm have written: "One man holds it (dibble) in his right hand and gives two or three sharp thrusts with the point. With the first, the hole is opened; with the second, the ground is broken and loosened, using the stick as a lever; with the third, the soil is further loosened, by working the stick from side to side several times. In the resulting cavity, grains of seed corn are dropped from the left hand, and, with the stick, the soil from the same hole is knocked lightly over the kernels" (Kelly and Palerm, 1950: 111).

> again my incense, I because the corn field was mine. I burned my incense. Then with that we were finished. We began to eat. "Let's eat." Now we will finish eating. "Let's eat in truth. Thanks to our Father God for the celebration of our food and drink," we said, "for the happiness of our food and drink for passing our days." In order that the animal herds of the god of the mountain would not do us evil, they are the raccoon, the birds, the duck, the parakeet, the magpie; we asked permission of the god of the mountain so that our corn would not be touched. Nothing, nothing would touch it. We finished the planting. Then late I went again to burn my incense, late in my corn field the second time. Then I left. I will not go back until seven days have passed.

Intercropping.—Though maize has traditionally been considered the basic source of food for the high Maya civilizations, some authorities are today questioning its primacy and are suggesting that root crops may have played a much more important role than is commonly supposed (Bronson, 1966). It is true that the potentiality of such crops has been consistently overlooked for the Maya. They are grown all over the area today, yet we have very little information on them; the bulk of the reports on lowland agriculture concentrate on maize.

Two of the most common root crops in the Maya lowlands are manioc (*Manihot esculenta* Crantz) and malanga (*Alocasia macrorhiza* Schott). Bronson, in an article calling attention to the possible role of roots in the prehistoric Maya lowlands, points out that six acres of manioc produces more than one-third as much food as seventy acres of corn. He then argues that if this 6:70 ratio between the areas planted to manioc and to maize had existed in sixteenth-century Yucatán, Spanish visitors may not even have noticed it, much less have recognized that the Yucatecans raised a third as much of it as of corn. He recognizes that manioc is the only basic root crop found in the lowlands today which does not occur there in wild form. Yet he counters that more than maize, manioc is the superlatively domesticated plant. It has been cultivated for so long that its ability to produce seeds has been entirely lost. No likely wild ancestor for it has ever been located (Bronson, 1966: 266, 269).

Malanga, grown as much or more than manioc in Chichipate today, is less productive than the latter. Nevertheless, on the basis of studies made in Africa, it has been shown to be four times as productive per unit of land as maize (Bronson, 1966: 269). Morley, who so extolled the virtues of maize, might have done well to look at such root crops.

In Chichipate, planting of intercropped cultivates assumes major importance as soon as maize has broken through the ground. Three root crops (manioc, malanga, and sweet potatoes), plantains, and bananas are the first things set out. Both types of manioc, white *(sak)* and yellow *(k?an)*, may be placed in the center as well as at the border of the milpa. For each plant, four hardwood cuttings, 8 to 10 inches long, are buried approximately 5 inches below the surface of the ground. About 2¼ *varas* are left between one planting and another. Set out in June, July, and August, manioc requires three months before it is harvestable. After the first harvest, it may continue to yield from two to a maximum of three years; if the milpero wishes to continue harvesting it, however, he must periodically clear away the competing brush.

Malanga planting is also an easy operation. The setting out of as many as two hundred plants will take only about seven hours of labor. One hour is needed for gathering the tubers, two for cutting them up, and four for planting. They are commonly placed both in the middle and at the margins of milpas.

The sweet potato is the only root crop to have a pan-Maya distribution and is easily the one most frequently mentioned in historical sources. According to Bronson, the thesis that this plant ((*Ipomoea batatas* (L.) Lam.)) is an American native is not seriously disputed at the present time. What is still in dispute is where in the Americas the plant was first domesticated. Since it is grown from cuttings and retains only a very limited self-propagating ability, it has probably diverged significantly from its wild ancestor (Bronson, 1966: 258).

In Chichipate, all three local varieties of sweet potatoes, white *(sak)*, red *(cak)*, and purple *(morado)* are sown together indiscriminately. Soft earth *(k?un ru li ch?och?)* is the preferred planting medium. The technique is simple: The earth is broken open with a machete, three to five slips are set in place, and from 1 to 2 *brazadas* away the operation is repeated. Sweet potato vines, if left alone, will soon cover the milpa floor, and the yield is abundant. However, sweet potatoes are not popular dietary items, so their potential is relatively untapped. Throughout the fallow milpas of Chichipate vast stretches are overrun with sweet potato vines, and no one bothers to dig the tubers.

Both sweet potatoes and manioc seem to share considerable antiquity in the Maya area. McQuown, for example (1964: 80), has presented a list of words which are held in common by all major branches of the Maya linguistic stock. On this list appear ten plant names: aguacute *(ˀoong)*, coyol *(meep)*, chili *(ˀik)*, cacao *(kakaw)*, ceiba *(ˀinup)*, calabaza amarilla *(hˀuXm)*, camote *(ˀi̋is)*, henequin *(kih)*, yuca *(tzˀiXn)*, and maiz *(iˀm, iˀx)*.[29] Of this list of plants, five are true domesticates: chili, yellow squash, maize,

A series of adjoining, intercropped wet season milpas. Visible are palm, chili, maize, and banana.

sweet potato, and manioc, and all five of these are basic cultivates in Chichipate today. The two root crops dovetail nicely with the non-root crops. Since root crops may be left in the ground indefinitely, there is no danger that their seed will spoil in storage before planting. They seem, furthermore, to be more drought resistant than maize (Bronson, 1966: 271).

Root crops are set out with a planting stick *(macān)*, a machete *(chˀīchˀ)*, or a combination of the two. The same implements are used for setting out plantains and bananas which are spaced about 3 *varas* apart. The three varieties of banana, green *(rax quenēy)*, white *(saki quenēy)*, and "stretched" *(rinoc quenēy)*, as well as the

29. The Maya base forms, given in parentheses, are from an article by Kaufmann which accompanied that of McQuown (Kaufmann, 1964).

three varieties of plantain, small *(tz'ul tul)*, red *(caki tul)*, and white *(saki tul)*, are all treated in identical manner. Collection of ripe fruit begins long after wet season maize has been harvested and continues for a maximum of two years, after which the milpa is allowed to revert entirely to secondary growth. During the period of production, milperos clear away the bush that grows up from time to time, and may even dibble more maize into the milpa for a dry season crop. In seeding near plantains and bananas, at least 1 *brazada* is left between them and hillocks of maize.

Other minor cultivates added to new wet season milpas include: pineapple *(ch'op)*, sugarcane *(utz'ajl)*, garlic *(ānx)*, onion *(cebolla)*, tobacco *(may)*, and peanut *(mani)*. None of these additions exact a great deal of planting time, and they are all set out either in small patches or at the margins of the milpas. However, the considerable variety of intercropped species, plus the house yard planting of such things as tomatoes *(pix)*, keep milperos fairly busy during the first weeks after seeding maize, squash, and beans.[30]

Fencing.—Once the milpa has been seeded, one of the most urgent tasks is to protect it from marauding animals. Fences are effective against domestic animal forms such as pigs and cattle, though they do little to keep out wild pests such as raccoons.

Fences last a maximum of three years, by which time they have practically rotted away. The only exception to this rule is the barbed wire fencing that the owner of Yuscarán has ordered strung around large sections of Chichipate, in anticipation of cattle raising in future years. Other than this exception, which really lies outside the indigenous milpa system, no fences mark off permanent boundaries. Where boundary marking is the sole object, a few upright poles are

30. Bronson has made a study of ancient Maya graphic art in an attempt to assess the dietary importance of intercropped items. He has found that this art represents maize frequently and roots never. He feels this does not negate the role of root crops, however. In his words: "Frequently . . . cultural esthetic emphases do not coincide with utilitarian, mundane importance. For instance, the Hanunóo emphasize rice farming to such an extent that religious observances at planting time apply to rice exclusively, and a man's status is determined by the amount of rice he can raise; yet, rice is not nutritionally very important, for only 37% of the Hanunóo's total carbohydrate intake is rice and maize combined (Conklin 1957). Yams and bananas, which the Hanunóo must have raised long before they began growing rice, provide the larger proportion of their caloric diet. No ritual attends their planting. Soustelle's Lacandón are similar in this regard. They have no rituals whatever for the planting of manioc, while corn-growing is characterized by intense religious feeling" (Bronson, 1966: 270).

set into the ground at crucial spots where two milpas adjoin *(sa*ˀ *xxucli nubal)*. When adjoining milpas are threatened by pigs, however, milperos may work together to construct a fence. Otherwise, one or the other may alone fence his milpa, even though part of the fence may lie along joint boundaries. Reciprocal labor is rarely used for fence building; if help is needed, it is usually recruited through the payment of wages.

Potential labor requirements for fencing are as great as those for clearing. A single *tarea* can exact a week of work if poles have to be brought any distance. Because of the amount of labor involved, shortcuts are used wherever feasible. In 1965 some fences in Chichipate extended over the plain only to the rise of nearby hills, in the hope that the steepness of the slope would discourage the incursion of pigs. Needless to say, this arrangement was somewhat less than 100 per cent effective, and by August numerous milperos were complaining about the damage pigs had caused.

Types of fencing.—In addition to the barbed wire recently introduced from Yuscarán, five basic types of fencing are used in Chichipate:

1. Open post fence for hogs *(xcoralil li āk)*. Posts used in building this fence are from 1 to 2 inches in diameter, and are tied together with vines. They are stuck upright into the ground, about 6 inches apart, to a depth of 12 inches. Their height averages around 4 feet and so, although the fence is built mainly for protection against hogs, it is also somewhat effective against cattle.

2. Solid rail fence for hogs *(xcoralil li āk)*. For constructing solid rail fences, milperos set sharpened poles 2 inches in diameter into the ground in pairs, with about 8 inches between one pole and another. They then drop cross poles into the slot left between each pair of uprights, building these cross poles up to the height of a man's chest. As soon as the uprights begin to spread apart, they are tied together with vines. Though a more substantial type of fence than the open post fence, the solid rail fence shares about the same degree of popularity.

3. Open rail fence for cattle *(xcorolil li cuacax)*. This construction, very similar to that of the closed rail fence, has the single advantage of requiring fewer cross pieces to build. Pairs of uprights are jabbed into the ground, and a few cross pieces separated by bits of logs are dropped between them. Though relatively ineffective for keeping hogs out of a milpa, it serves as an admirable barrier against cattle.

4. Brush fence for cattle *(xcoralil li cuacax)*. The most casually assembled fence of all, a good brush fence may be made only where there has been heavy forest cover, preferably a site which has just been cut over from primary or mature secondary growth *(nīnk li q'uiche')*. It is built by piling together logs and brush along the edge of the milpa to a height of up to 5 feet. Some of the trees used in assembling this fence may have a diameter of as much as 25 inches. A good brush fence, though put together in such a helter-skelter way, is quite effective in keeping both cattle and hogs out of fields.

5. Chicken fence *(xcoralil li caxlan)*. Chicken fences are constructed principally in house yards in order to keep fowl out of small garden plots. No posts are used. Instead, sapling poles from 1 to 1¼ inches in diameter are cut, sharpened, and stuck from 6 to 8 inches into the ground. Generally not more than 4 inches are left between one upright and another. For strength, the fence is lashed together with vines. It provides little protection against hogs. Since it is intended for protection against only chickens, the uprights are left to heights varying from about 10 inches to 3 feet or more. Pragmatic rather than aesthetic considerations prevail in the construction of this, as well as other fencing.

The greatest fear which drives people to build fences is that hogs will root out maturing crops. If this occurs, one milpero may open litigation against another, carrying the case to municipal authorities in El Estor. Should the owner of the marauding hogs not pay the fine imposed, his hogs may be taken as pawns and not released until the debt is canceled.

Most milperos, rather than invest extra time and effort in fence-building, and rather than risk quarrels with neighbors, prefer simply to make their milpas fairly distant from the village. A ten-minute brisk walk is considered minimum.

Watching.—In 1956-57 monkeys were practically exterminated in northern Central America by the great epidemic of jungle yellow fever that swept the area (Stuart, 1964: 317-18). Other animal pests remain, however, and cause crop damage no matter how tight the fencing. The most destructive is the *aj-ou,* or raccoon (*Procyon lotor shufeldti* Nelson and Goldman).[31] He begins his marauding

31. In Murciélagos, the Hempstead estate on the shores of nearby Lake Izabal, deer (*Odocoileus virginiana nelsoni* Merriam and *Mazama americana cerasina* Hollister) do more damage than do raccoons. No such reports were given in Chichipate, however. Other mammals which probably cause consid-

activity as soon as maize seed has been dibbled into the ground, breaking open fresh hillocks and carting away the grain. When the young plants have sprouted, and up to the time they pass a foot in height, he will break them over "from sheer meanness," as the milperos say. But it is during the period when ears are maturing that the raccoon's activities become the most pernicious. He will steal enormous quantities of grain. Where he cannot reach the ear, he breaks over the maize stalk. What is the most exasperating for milperos is that often this animal breaks over maize, tears out the ears, and simply scatters the grain over the milpa floor. A homeopathic magical belief has it that if a man breaks milpa ground with his hands, within three days the raccoon will damage his maize. At planting time, and during the maturation period, ritual prayers solicit protection against raccoons.

Other mammals which cause damage are the *chʔo* (*Sigmodon hispidus* Say + Ord. sp.), the *qʔuicheʔ āk* (*Tayassu tajacu nelsoni* Goldman), the *halau* (*Agouti paca nelsoni* Goldman), the *ākam* (*Dasyprocta punctata punctata* Gray), and the *sis* (*Didelphis marsupialis tabascensis* J. A. Allen). Of these the common pouch rat (*Sigmodon* sp.) is the most bothersome. Some damage also results from flocks of birds. The most common of these pests are the *puyuchʔ* (*Aratinga astec astec* Souancé) and the *patux* (unidentified duck). When wild fowl become a serious problem, a situation more likely to arise in dry than in wet season milpa, scarecrows (*aj xeʔbesihom cʔal*—"to frighten in the corn field") are made out of old clothes, rags, or paper, and set in those spots most affected.

No traps are made. When ritual fails, the only way of ridding the milpa of animal pests is to stay on watch and frighten or shoot them when they appear. Throughout the growing season, and particularly when there are signs of damage by marauders or when the maize is readying for harvest, milperos go to their fields at dusk, sleep in little lean-tos *(papāx)* of poles and palm fronds, and wait for the animals to arrive. When they hear noise in some part of the milpa, they walk slowly in that direction, light up the spot with a miner's lamp, and attempt either to shoot the culprits with a shotgun or

erable damage to growing maize, though seldom mentioned by milperos, include the dwarf opossum (*Marmosa m. mexicana* Merriam), the orange opossum (*Caluromys derbianus fervidus* Thomas), the prehensile-tailed porcupine (*Coendori m. mexicanus* Kerr), the coati (*Nasua narica narica* L.), the kinkajou (*Potos flavus campechensis* Nelson and Goldman), and the tapir (*Tapirus bairdii* Gill).

slaughter them with a machete. They take the meat home when they return at 4:00 or 5:00 in the morning. A milpero will average two to three nights of watching per week during the season when grain is maturing. If, however, an opportunity presents itself to make money working as a day laborer in the neighboring estates or in road construction, men leave their milpas to luck. "Let them eat what they want to; there will be plenty left for us" seems to be the prevailing attitude.

Weeding.—One of the advantages migrants find in their new environment is that weeding is much less a problem than it was in Alta Verapaz. This is due to the thoroughness with which fire clears over the land, given the density of vegetation in the lowland region, and is also due to the fact that Alta Verapaz has long been under intensive cultivation, a situation which in itself will favor large weed populations. With many weed species, repeated firing will promote rather than retard seed germination.[32]

In the highlands, five weedings are needed throughout the growing season. But in Chichipate, two weedings and one cutting back of sprouts from sapling and tree stumps is the maximum care that one must give. Indeed, in milpas planted on land just cleared of mature growth, a single lopping off of sprouts may be all the weeding that has to be done—hence the strong theoretical preference for this type of site.

Of the ligneous growth that tends to sprout back immediately after firing, three forms are the most common: *cakaj* ((*Bursera simaruba* (L.) Sarg.)), *pōk* (*Spondias purpurea* L.), and *ak*ʔ*l* (Moraceae). With herbaceous materials, grasses and sedges are the worst pests. These are divided into only two general categories, stoloniferous *(pach*ʔ*aya*ʔ*)* and non-stoloniferous *(q*ʔ*uim)*. Each represents, however, various plant types in the European binomial system, as the following listing indicates.

*Pach*ʔ*aya*ʔ:
- *Paspalum conjugatum* Bergius
- *Axonopus compressus* (Swartz) Beauv.

*Q*ʔ*uim*:
- *Scleria pterota* Presl.
- *Panicum polygonatum* Schrad.
- *Paspalum caespitosum* Flugge
- *Paspalum botterii* (Fourn.) Chase

32. For detailed weed information for Yucatán, cf. Steggerda, 1941: 99.

Setaria geniculata (Lam.) Beauv.
Digitaria horizontalis Willd.
Cyperus eggersii Boeckel.) = sedges
Cyperus odoratus L.)

Of the other herbaceous pests, vulture bean (*soʔsol quēnkʔ—Cassia occidentalis* L.) is the most bothersome. Ligneous growth which creates minor problems includes *yaxteʔ* (*Brunfelsia* sp.), *kʔinam* ((*Vatairea lundellii* (Standl.) Killip ex Record)), *hu* ((*Virola guatemalensis* (Hemsl.) Warb.)), *rosul* ((*Swartzia guatemalensis* (Donn. Smith) Pittier)), *guamo* ((*Tabebuia pentaphylla* (L.) Hemsl.)), and *kʔerk* (*Heliconia latispatha* Benth.).

There are a considerable number of useful plants which, once the milpa is established, may be treated as semi-domesticates. Only where they present a major threat to maize will they be entirely cleared away. The most common of these are listed in Table 14.

Weeding techniques.—Milperos in Chichipate recognize three basic purposes in weeding their fields: to keep other plants from competing with maize, to keep the milpa floor moist by finely chopping the growth they have slashed or uprooted, and to clear away spots for the planting of intercropped cultivates. They do not normally associate a heavy stand of weeds with increased depredation by animal pests.

The techniques used in freeing the milpa of weedy growth are classed as three:

1. *Raplēnc*—to cut back sprouts from tree stumps. If the milpero is right-handed, this is done with the machete in the right hand, using swift, downward strokes.

2. *Xmichʔbal li pim*—to uproot weeds. This is one of the few milpa tasks in which women take a major part. The operation consists of leaning over, grabbing the plant in one or both hands, and pulling it up, roots and all The uprooted plant is left to dry on the milpa floor.[33]

3. *Akʔinc*—to clear with a hoe. The tool used is the European variety of hoe; it is held in both hands, and with it weeds are both uprooted and chopped.

33. Judging on the basis of aboriginal tool types, Morley considered uprooting to be the only weeding method practiced in pre-Conquest times. This method was used on an experimental plot in Yucatán, after cutting had been practiced for four previous years. The year that uprooting was substituted for weed cutting, the yield slightly exceeded even the first year's crop. The yield was cut to less than half the next year (the sixth year), gained the seventh

Sprout cutting *(raplēnc)* is normally performed only once, anywhere from four to eight weeks after maize planting. Milperos, however, carry their machetes with them wherever they go; consequently, they are constantly hacking at obtrusive sprouts whenever

TABLE 14
COMMON SEMI-DOMESTICATES

Semi-domesticates	Use
Macuy (*Solanum nigrum* L.)	Leaf is eaten boiled with chili.
Sal tul ((*Pouteria mammosa* (L.) Cronquist))	Ripe fruit is eaten raw.
Camānc ((*Neurolaena lobata* (L.) R. Br.))	Herb used to treat stomach ailments.
Piak (*Dioscorea* sp.)	Boiled root eaten seasoned with salt.
*C*ʔ*ante*ʔ ((*Gliricidia sepium* (Jacq.) Steud.))	Cold water solution containing crushed leaves used in bathing a patient to combat fever.
Mocōch ((Orbignya cohune (Mart.) Dahlgren ex Standl.))	Fronds used for roofing, fruit eaten, and oil used in making soap.
Chiltep (*Capsicum frutescens* L.)	Wild chili pepper, used same as domestic chili pepper.
Obel (*Piper umbellatum* L.)	Condiment for fish.
Mox (Marantaceae)	Fresh leaf used for wrapping meats and other bundles.
*Chi*ʔ ((*Byrsonima crassifolia* (L.) HBK))	Ripe fruit eaten raw.
Pec imis (*Thevetia* sp.?)	Ripe fruit eaten raw.
*Cuāch*ʔ*il* ((*Dialium guianense* (Aubl.) Sandw.))	Ripe fruit eaten raw.
Ses (*Amaranthus hybridus* L.)	Leaves eaten boiled.

they walk through their milpa. Uprooting *(xmich*ʔ*bal li pim)* is preferred, by far, as a method of clearing away heavy herbaceous growth. It is practiced constantly, though a major effort of this sort is seldom made more than twice during the growing cycle.

year, and dropped to almost nothing the eighth year because of a plague of locusts. On the basis of these figures, Morley concluded that the ancient method of weeding (uprooting), while it may not have produced more corn in any one year than the modern method (cutting and hoeing), prolonged the life of the average cornfield from two to perhaps seven or eight years before it finally had to be abandoned (Morley, 1947: 152-53).

When weeds or grasses become especially burdensome, most milperos, rather than spending days clearing them away, prefer to seed velvet bean *(quẽnk' caballo)* and let it take over the milpa floor. As it does so, it shades the grasses until they die, but still permits maize, manioc, and plantains to grow to maturity. Hoeing (*akˀĩnc*—"cleaning") is of much less importance in the lowlands than in the highlands of Guatemala. Milperos who hoed their maize fields five times a season in Alta Verapaz have found that a maximum of two times is sufficient in Chichipate. In milpas that have just been carved out of mature forest, not even this much is required.[34]

Milperos who reported man-hours spent in weeding in 1965 were averaging eighty-four per *manzana*. The lowest number of hours was twenty-four, for a large milpa of 33 *tareas*. The majority of these *tareas,* twenty-four, were planted in a site freshly cut from mature growth, however, so all the time spent in weeding had been in the nine remaining *tareas* which were planted in what, before clearing, had been low secondary growth *(cocˀ pim)*.

Disease and insect pests.—Diseases are quite prevalent in older milpas. Chichipateños have no means for coping with them, and when they become serious, the field is simply abandoned.

Serious damage also results from the activity of insects. These may attack crops at all stages of development, as well as the contents of granaries once maize has been harvested. Chichipateños deal with them basically through supernatural means, insecticides being fairly inaccessible and unknown. When insect activity becomes a major problem, milperos go into their fields and collect all the specimens they can find in *huacales (jõm)*. These they place before the household shrine as an offering to their patron saint. After burning candles and copal, and pleading for protection against further infestation, they wrap the pests in *moxan* leaves, place them on the cooking fire until they are reduced to ashes, and finally scatter the ashes in the corners of the milpa.

34. Chichipate data correspond with what Bronson has said about the differential use of the hoe in the Guatemalan highlands and lowlands (Bronson, 1966: 255). Sanders (1962) has argued that the rare use of hoes in the Maya lowlands constitutes a critical difference between Nigerian and Middle American farming systems: without hoes, grass is more likely to become established and the milpas rendered useless. One report for the Petén, however, states that farmers there regularly make farms on the grassy savanna, using "hoe-like" sticks to turn the turf (U. Cowgill and Hutchinson, 1963b: 277).

Some of the milperos feel that insect activity is greater in older than in newer milpas. This would seem to correspond with Ursula Cowgill's experience with plots in the Petén. There, no first year plot examined had corn borer or any other form of insect damage, with the exception of a small amount caused by colonies of leaf cutting ants. Both corn borer and leaf cutting ants were prevalent, however, in second year plots (U. Cowgill, 1962: 283).

A few magical taboos are observed to protect crops from insects. For three to nine days before and after rituals performed at various points in the milpa cycle, milperos neither eat fruit nor sleep with their wives. The belief is that if they were to do so, worms would destroy their crops.

Insecticides are used only for protecting stored maize against weevils. As each layer of ears is set out in the granary, water is sprinkled over it, and then some milperos add a fine dust of Gamaxan (gamma isomer of benzene hexachloride plus DDT). When weevils attack stored maize, kernels are quickly perforated and their cores converted to a fine and worthless dust.

Thirteen arthropods constitute major threats to production in Chichipate (Table 15). Of these, five (indicated by a dagger) attack maize while it is standing in the milpa, and one (indicated by an asterisk) after it has been harvested and stored. The rest concentrate on intercropped and house yard cultivates other than maize.

Growing ritual.—Except for the offering to the saints of worms collected in the milpa, growing ritual is limited to ceremonies conducted when tassels are just ready to emerge on the maize (*bōloc*—literally "rounded"). These ceremonies may begin at any time during the day, with the milpero and his wife burning a candle and copal for the household saint. They then walk to the milpa and kneel in the heart of the field *(xyil)*, e.g., at the spot where the ritual cross stands. They pray and shake copal three times toward the cross, then three times to the east, three to the north, three to the west, and finally three to the south, placating in this way the four corners of the field. After incense has been offered to these four corners, they then circle with the brazier around the heart of the milpa three times. Finally, the contents of a bottle of chicken blood are tossed as a libation, first to the right, then to the center, and then to the left rear of the milpa. This is followed by a simple offering of libations of bitter chocolate. After the libations, prayers are lifted to God, Jesus Christ, Mary, the Holy Spirit, the angel

Jehovah, Saint Rafael, Saint Michael, and the spirits of the hills, that the maize may grow evenly and full. The ceremony is terminated with the Lord's Prayer.

Once the milpa ceremony has terminated, the couple return home and make identical prayers before their household shrine. As they

TABLE 15
COMMON ARTHROPOD PESTS

K'ani c'an	Coleoptera: Chrysomelidae; *Cerotoma* sp. and *Oedionychus* sp. Eats leaves of manioc, beans, malanga, watermelon, papaya, chili pepper, chick pea, and tomato. Especially active in August.
K'eki c'an	Coleoptera: Chrysomelidae; *Lema* sp. and *Nodonota* sp. Eats beans, malanga, watermelon, papaya, manioc, chili pepper, chick pea, tomato. Active in August.
Motzo'†	Dermaptera: Forficulidae; *Doru lineare* (Esch.). Eats maize stalks and ears, leaves, tassels, and silk. Also eats malanga, papaya, watermelon, sugarcane, squash, sweet potato leaf, and tomato. Active in November, December, May, June, and July.
K'ani motzo'	Diptera: Otitidae. Enters squash stalk and "cuts off the circulation." Tips of the plant then die. Active all months of the year.
Tequen†	Formicidae: *Atta* sp. Attacks the leaves of maize, watermelon, fruits. Active all months of the year.
*Moc'oc'**	Coleoptera: Curculionidae; *Sitophilus oryzae* (L.) Attacks stored, dried maize kernels. Leaves only shell and whitish powder. Active all months of the year.
Caki sānc	Formicidae: *Solenopsis* sp. Attacks garden vegetables and fruits. Active all months of the year.
K'eki sānc†	Formicidae: *Solenopsis geminata* (F.) Eats maturing maize ears, as well as tassels. Attacks malanga, sugarcane, sweet potato leaf, manioc, chili pepper, tomato, tobacco, squash. Active all months of the year.
K'eki motzo'	Noctuidae: *Spodoptera frugiperda*. Attacks beans, malanga, and garden vegetables. Bores through ripening pineapples. Active all year.
K'eki chili'†	Orthoptera: Acrididae; *Orphulella* sp. Cuts maize stalk when it is less than 5 inches high.
Gusān	Hemiptera: Pentatomidae. Attacks tomato plants.
Cojoj	Diplopoda: Spinabolidae; *Orthoporus* sp. Eats leaves and tender fruits of garden vegetables.
Sāc†	Orthoptera: Acrididae; *Ambyltropidia* sp. Eats tender, green leaves of any plant.

*Attack maize after it has been harvested and stored.
†Attack maize while it is standing in the milpa.

do this, they burn candles and copal and offer the saint food in the form of chicken broth, tortillas, tamales, and bitter chocolate. Again, the food is left on the shrine until late evening, when children are invited to eat it. For three days before and three days after the ceremony, both the milpero and his wife remain sexually continent, and everyone in the family refrains from eating fruit.

House yard gardens.—Aside from spending time and effort in intercropping activities during the period when maize is maturing, milperos engage in limited projects of planting in house yards. These sites, because of their convenience and small size, are the best for growing items marginal to the diet. Lime trees *(lamunx),* pigeon pea *(che? quēnk?),* sugarcane *(utz?ajl),* tomatoes *(pix),* various varieties of squash *(c?um),* papaya *(papaya),* bananas *(quenēy),* and sweet potatoes *(is)* are the most common of the marginal cultivates. In addition, new items for general intercropping are customarily tried first in small house yard plantings. The extent of these house yard plantings, whether marginal or experimental, is limited by the need to tightly fence everything that might be damaged by the rooting of hogs.

Harvest (k?olbil).—As in farming communities in all parts of the world, the harvest period in Chichipate is a time of increased, and usually joyous activity. From the time seeds have been dibbled into the ground, milperos have observed each stage of development with great interest. As the first ears matured they carried them as gifts to friends, neighbors, and relatives to show that they now had the first fruits of harvest.

The principal stages designated in maize development are thirteen: 1. *āubil li c?al*—dibbled into the milpa, 2. *jutz?ju li coc? cuaj*—broken through the ground, 3. *jun xben rak*—knee high, 4. *jun sa?*—waist high, 5. *cham li c?al*—thick, cannot see through the field, 6. *bōloc*—tassels emerging, 7. *aj li c?al*—grains appearing in the ears, 8. *cuan li och*—there are ears, 9. *cuan lin c?ux*—there are ears for eating, 10. *x?oc xk?anal li c?al*—milpa is beginning to ripen,[35] 11. *x?oc xchakial li c?al*—milpa is beginning to dry, 12. *anakcuan chaki li c?al*—milpa is now dry, ready for harvest, 13. *k?olbil*—harvested.

With early varieties of maize, soft-grained ears may be gathered in small quantities as early as the last days of July. In this form, they are never used as a staple, but only as a side dish. The real harvest,

35. It is interesting to note that the word *k?an* has the double meaning of /yellow/ and /mature/, indicating the central place of maize for all cropping.

when dried ears are plucked on a large scale, does not begin until after August 15. Only early varieties, such as black maize *(kʔeki hal)*, yellow maize *(kʔani hal)*, and small white maize *(cocʔ saki hal)*, are available at this time, however. The harvest of large white maize, the main staple of the community, cannot begin until a

An intercropped house yard. Plants visible are sweet potato, chili, squash, banana, sugarcane, pigeon pea, and manioc.

month later, and often the month of October becomes the busiest of the entire harvest period. By November 1 practically all harvest activities have been completed, and efforts are being transferred to the preparation of dry season milpas. Should a choice have to be made, it is better to let maize stand a little too long than to harvest it prematurely. If the grain is soft when picked, it will rot in the granary.

Harvest groups are made up of both hired and reciprocal labor. In the case of ten of these groups in 1965, the size ranged from one to fifteen, with a mean of 4.7. The number of man-hours required per *manzana* ranged from 60 to 137, with the mean being 106. This mean is derived from the time involved in harvesting alone; it does not include granary building, carrying, storing, or shelling.

The purported advantage in forming large groups for harvesting is that when maize becomes very dry, the kernels loosen and the crop is thus much more susceptible to predators. The milpero of Chichipate wants to bring his harvest in as close to the right time as possible, and for this reason he forms work teams just as does the wheat farmer of the traditional Midwest.

No divination seems to be associated with the setting of a date for harvest, though informants state that it is preferable to harvest from three to ten days after the appearance of a "sleepy moon." By "sleepy moon" they are referring to the full moon that rises about 7:00 P.M. Harvest activities may be carried out both on sunny and rainy days. Rainy days are preferred, however, for maize harvested then is more resistant to the weevils that attack the granaries. If maize is harvested on a sunny day, it must be sprinkled with water before being stored, as a precautionary measure.

Green maize *(aj)* is harvested in sizeable quantity only on the first day that a milpero finds that it is edible. This is the time he shares it with neighbors and friends. From then on throughout the harvest season, families continue to rely principally on old stores of dried maize. A few green ears are brought only every four or five days and used simply as a side dish.

Harvest ritual.—Ritual proscriptions dictate that green maize brought in before the appearance of the "sleepy moon" come not from the principal milpas, but rather from small marginal parcels that have been planted to early types. To bring maize from the main milpas, one must wait for the "third sleep" of the moon. On the first night or first sleep *(xbēn xcuara li po)*, it rises about 7:00 P.M. On the second night or second sleep *(xcab xcuara li po)*, it rises a little later, and on the third *(rox xcuara li po)*, still later. Only after *rox xcuara li po* can one safely bring in green maize from his main milpa. If brought in before this, it is believed that the entire crop, once stored, will be eaten by weevils. In 1965 the three "sleeps" of the moon came on August 12, 13, and 14.

For the first real harvest of green corn, as much as one hundred ears may be cut. As soon as these are brought home, the woman of the house begins to grind them for *atole*. Three *huacales* of *atole* are then presented to the saint, along with tortillas, tamales, and bitter chocolate. Upon presentation, these are fumigated with copal. Candles are burned on the shrine and prayers are led by the household head. The following is an example:

Kekchi	English
Ay Dios, anakcuan xeˀoc chak lin cua li cuucˀa haˀ. Oc cue xkˀolbal. Xin-oc chak, xinchˀutub saˀ li rochochil. Anakcuan xebinmol li cuas li cuītzˀin saˀ xcuanquil lin hal nak xqˀue xcut chiku li lokˀlaj tzūl lokˀlaj takˀa. Bantiox a cuaˀ lokˀlaj tzūl lokˀlaj takˀa. Anakcuan xaqˀue xsahil xchˀōl lin cua cuucˀa haˀ. Junab chic, joˀcan ta cuiˀchic, ay Dios a cuaˀ. Anakcuan bantiox saˀ scˀubaˀ li Dios Acuaˀbej, li Dios Cˀajolbej, li Dios Espiritu Santo amen. Saˀ xcˀabaˀ li Dios Acuaˀbej, li Dios Cˀajolbej, li Dios Espiritu Santo, amen. Saˀ xcˀabaˀ li Dios Acuaˀbej, li Dios Cˀajolbej, li Dios Espiritu Santo, amen.	Oh, God, now they brought in my food and my drink. I'm beginning to harvest. I begin, I gather it in the granary. Now I have gathered together my fellow men in the power of my ears of corn that the holy god of the mountain gave to me. Thanks, father, holy mountain, holy below. Now you have given the happiness of my food and drink. One year more, like this again, oh God, father. Now thanks in the name of God the Father, God the Son, God the Holy Spirit, amen. In the name of God the Father, God the Son, God the Holy Spirit, amen. In the name of God the Father, God the Son, God the Holy Spirit, amen.

Once the prayers have finished, children are sent to present the cooked green corn to neighbors and relatives. In a spirit of great rejoicing, they announce: "I want you to taste my things." "Ah, have you now begun?" ask the recipients. "Ah, blessed be God."

In this as well as in other harvest ritual, involvement tends to be limited to household members only. Relations between the supernatural and man are very localistic, and community-wide ritual occurs on only the two saint's day celebrations of George and Dominic. At that time, all-night drinking and dancing overcome ordinary social reticence.

Household members perform no further harvest ritual until the day the granary is filled and set in order. During the noon lunch break, the incense brazier is brought, and the little structure is fumigated by moving the brazier in a circle. A candle is placed in the half of the granary left empty for sleeping, resting, and watching. A prayer is lifted to the spirit of the maize:

Kekchi

Ay Dios, x-oc chak inhal. Anakcuan a cuaʔ xeʔx-oc chak chijunil. Xeʔx-oc chak saʔ li rochochil. Cuaʔe. Mexcʔoxlac. Anakcuan saso saʔ lē chʔōl, saso saʔ lē ranima. Anakcuan xic ke chacʔambal. Māre cuan xcana chak jun, cuib saʔ li rok cuaj, pero teʔchālk lix muel. Anakcuan tinxic xxocbal. Quimquex a cuaʔ. Quimkex anakcuan. Texchālk. Texincʔat, tinbok, texinbok riqʔuin utzʔuʔuj. Li tzūl takʔa tambien tixcokʔ chak lē muel. Toj saʔ junab chic nequeʔxic ecaʔ cua. Pero moco arin chic. Māre jalan chic aj iqʔuin. Saʔ xcʔabaʔ li Dios Acuaʔbej, li Dios Cʔajolbej, li Dios Espiritu Santo, amen. Saʔ xcʔabaʔ li Dios Acuaʔbej, li Dios Cʔajolbej, li Dios Espiritu Santo, amen. Saʔ xcʔabaʔ li Dios Acuaʔbej, li Dios Cʔajolbej, li Dios Espiritu Santo, amen.

English

Oh, God, my ears of corn have entered. Now Father, they all have entered. They have entered in the granary. They are there. Don't you all worry. Now your hearts are happy. Your souls are happy. Now we are going to take you. Perhaps there are one or two left on the stalks, but their spirits will come. Now I will go to gather them. All of you come, Father. All of you come now. You will come. I will burn you all. I will call. I will call you with flowers (incense). The god of the mountain will release your spirits. Until one more year they go your second time. But not here again. Perhaps with a different one. In the name of God the Father, God the Son, God the Holy Spirit, amen. In the name of God the Father, God the Son, God the Holy Spirit, amen. In the name of God the Father, God the Son, God the Holy Spirit, amen.

Lunch is eaten, and the workers return to the field. At the end of the day the household head and his wife once again burn incense for and lift prayers to the spirits. The mood is one of thanksgiving. One milpero's prayer on this occasion refers to ears of corn as his children:

Kekchi

Ay, Dios, xacuocsi chak xmay lin hal. Anakcuan a cuaʔ tincʔux tincuaʔ lin cua re xsahil inchʔōl. X-ocsi incua li cuucʔa haʔ saʔ rochochil. Xchal lin cʔulaʔal. X-oc saʔ li rochochil lix muel lin hal. Anakcuan naxic cuiʔchic saʔ junab. Tinpatzʔ cuiʔchic chacue chacuu at cuaʔ. Anakcuan tintzʔak li bendicion saʔ xben nak saʔ xcʔabaʔ li Dios Acuaʔbej, li Dios Cʔajolbej, li Dios Espiritu Santo, amen. Saʔ xcʔabaʔ li Dios Acuaʔbej, li Dios Cʔajolbej, li Dios Espiritu Santo, amen. Saʔ xcʔabaʔ li Dios Acuaʔbej, li Dios Cʔajolbej, li Dios Espiritu Santo, amen.

English

Oh, God, you brought in the spirit of my ears of corn. Now, father, I will eat my food for my heart's happiness. My food and drink have been brought into the granary. My babies have come. The spirit of my ears of corn has entered the granary. Now it goes again another year. I ask again from you, Father. Now I partake in the blessing on it in the name of God the Father, God the Son, God the Holy Spirit, amen. In the name of God the Father, God the Son, God the Holy Spirit, amen. In the name of God the Father, God the Son, God the Holy Spirit, amen.

Implicit in these prayers, as well as in the Chichipateños' general attitude toward milpa farming, is the animistic concept that natural phenomena are, in effect, sentient spirits. God is asked to be with the maize so that it will not be sad, even though it has been accidentally left in the field, and even though it may be eaten by wild animals. Excessive loss through careless harvesting is countermanded by magically calling all the maize that has been left into the protection of the granary.

The ritual of harvest completion begins around noon and continues until 7:00 or later in the evening. In addition to prayers, offering of candles, and fumigation with copal, it involves libations. These may be of chicken or turkey blood, and usually also include bitter chocolate. Once the spirit of maize has been properly fed and the protection of heavenly spirits has been dutifully invoked, the family returns home and there offers similar homage to the household saint. Burning copal and candles, *huacales* of chicken broth and bitter chocolate, tamales, and tortillas—the standard offerings—are all placed on the family altar. As on other ritual occasions, these are left there until thoroughly cooled, at which time the children of the house are asked to feast on them.

Granary building.—New granaries *(rochochil li hal)* must be built every year. This is not only because milpa sites are constantly changing, but also because the construction is so fragile and insect pests so ubiquitous that the structure seldom lasts longer than one to two years.

The size of the granary varies with the size of the milpa and the expected size of harvest. As a general rule, the building is placed at the very entrance to the milpa, as close as possible to the hamlet. Storing maize there, rather than in one's house, has three purposes. First, it postpones the task of carrying to a time when the milpero is less occupied than he is at the time of harvest. Second, from bitter experience, Chichipateños have learned that destruction by rats is much worse in a concentrated settlement than it is in the open field. Third, they believe that the greatly feared fer-de-lance, since it eats rats, will be more attracted to human settlement if large stores of grain are located there. Within the hamlet, there are only two ways in which some degree of protection may be attained from predators: one is to hang bunches of maize ears from rafters in the center of the hut, and the other is to store the grain, usually shelled, in a tightly closed coffer. Only a limited supply may be stored in this fashion, however, because of the limited number of rafters and the rapidity with which weevils invade stores of shelled maize. In spite of all their precautions, milperos state that rats may eat up to ten ears per household per night, and that shelled maize may never be kept longer than one month, no matter how tight one's coffer is.[36]

Outside labor is very rarely recruited for granary building. The size of construction parties in 1965 ranged from one to four, all these numbers being recruited from the household itself and lone workers

Maize stored in a granary.

prevailing. The average time employed in granary construction was 70.9 man-hours per unit; the greatest was 216 hours for a large household that farmed 7 *manzanas* of milpa.

The mechanics of constructing a granary are relatively simple. Sapling-sized poles are felled in the bush with a machete and gathered together at the entrance to the milpa. This material is supplemented with palm fronds and vines. No foundation is laid; rather, a framework of heavy poles is first erected to create skeletal walls. Rafters are tied to this in such a way as to create a pitched roof. Once the framework has been completed, the rest is relatively easy. Except for two large openings at the sides, walls are filled in by

36. Almost identical methods of storage have been recorded by Steggerda for Pisté, Yucatán (Steggerda, 1941: 109); by Redfield for Chan Kom, Yucatán (Redfield and Rojas, 1962: 45); and by Kelly and Palerm for the Tajín Totonac (Kelly and Palerm, 1950: 117).

setting sapling-sized poles upright into the ground, leaving as little space between them as possible. The roof is finished with a thatch of overlapping palm fronds, in a similar though more slipshod manner than that used in house construction in the village. Finally, a floor is constructed by laying saplings side by side until the entire interior is covered. On top of these saplings is placed a heavy, tight layer of palm fronds which raises the level of the floor some 5-7 inches off the ground, the purpose being to prevent rotting of the stored maize.

As a final precaution against predators, many milperos clear away the bush for up to 50 yards in all directions from the granary. In this space, they plant cultivates such as chili peppers, tomatoes, watermelon, garlic, onions, and beans so that, with needed tillage, the bush will not grow back during the year.

Harvesting procedure.—On commercialized estates in northern Guatemala, each group of harvesters tends to be commanded by a "coronel." Among Chichipateños, such highly organized cooperation among harvesters does not exist. Each worker goes indiscriminately through the field, without attempting to follow rows (which hardly exist anyway) and without even having a formally divided section in which to work. Thus, he acts not unlike the milpero who harvests alone *(kʔoloc xjunes)*. Custom does dictate, however, that workers begin on the side of the field opposite to the granary, and usually they simply line up along that border and then work irregularly toward the other side. On sloping land, they work from the top down.

Harvesting equipment is exceedingly simple, consisting of nothing more than the worker's hands, gunny sacks purchased from truckers, and a tumpline *(tāb)*. Women remain home and prepare food for the harvesters, who expect to be provided with pork or beef, as well as with bitter chocolate, tortillas, tamales, broth, and *atole*. Three large pieces of meat are customarily considered minimum for each man, and the host provides *moxan* leaves for wrapping so that the workers may take home any meat which is left over at the end of the day.

A number of reports on lowland Maya agriculture have emphasized that maize stalks are bent over when the plant has grown and the ear begun to develop. Reina (1967: 5) claims that this is done "to allow the ear of corn to mature properly." Steggerda (1941: 108) claimed that it was done "to keep the rain from entering the ear and spoiling the grain through mold," or perhaps so that "birds have less chance of tearing open the husk." Hester (1954: 162) prefers the theory of protection from rain. Chichipateños, however, leave the stalks untouched until the time of harvest itself. Then, because of the height of the maize (some stalks are over 12 feet high), they are forced to bend it over. In doing so, they grab the maize stalk in the left hand about 1½ meters above the ground. Holding the bent stalk with the left hand, they then break back the ear with the right until it is severed. They shear off two or three layers of husks, and then throw the harvested ear onto a pile that slowly builds up in the middle of the section on which they are working.

At least one of these piles or mounds is developed per worker each day. In a milpa of average yield, it will grow to about five large gunny sacks of unhusked ears, and this normally represents a *tarea* or day's work. At the end of the day, the ears are put into bags and taken to the granary. There they are dumped without order in the middle of the floor. On finishing the day, the host asks his helpers whether they want money or maize. If they prefer money, he pays them on the spot. If maize, he counts out 100 ears for each worker who puts them into a borrowed gunny sack, carries them home, and returns the sack to the host later in the evening. In poor milpas, the usual custom is to pay with money or reciprocal labor rather than maize. A good 1-*manzana* milpa should yield up to 32 *fanegas* (400 ears each) of unhusked maize which, when shelled, will result in around 20 *quintales* of grain.

During the harvest season, milperos and their families often sleep in the granary, especially if the milpa is a considerable distance from the village. The watching hut, where the milpero formerly spent occasional nights, has usually reached a sad state of repair by this time and is simply allowed to decay. Once built, the granary may therefore be used in watching against marauders. It is especially helpful in this regard during the period of dry season recultivation.

Because of the cost involved (each worker receiving either 50 cents or 100 ears of maize for a *tarea* of work), most milperos are reluctant to bring in harvest helpers for more than one day. This is normally done to begin the harvest of the most marketable and widely grown variety, large white maize *(nīnk saki hal)*. From that day until the end of harvest, a man works either alone or strictly with members of his basic milpa labor group, i.e., with those individuals who share his granary. While hired helpers are usually relatives or close friends, it sometimes happens that individuals come from Alta Verapaz looking for day labor in the lowlands and are given work. Children become eligible for harvest labor at age eleven.

Carrying and storing.—The basic tool for carrying, the tumpline *(tāb)*, is made of balsa *(puj)* bark. With only a tumpline, gunny sacks, and the sweat of their brows, milperos gradually build up a large, disorderly pile of maize in the middle of their granaries. Every two to three days they are forced to stop and put things in order. Classing the ears by quality is the first step taken. The smallest ears, those which have not filled out completely, and those eaten

Carrying grain from storehouse to hamlet.

by worms are set apart on one side of the granary for domestic use during the first weeks after harvest. They may be used as chicken, turkey, and hog food as long as they last. This inferior maize *(xsaˀ li hal)* is simply tossed in a disordered heap on one side of the granary; both its quality and its disorderliness offer a striking contrast to the neat, rich layers of ears that are then carefully built up on the other side.

The best and largest of these ears *(nīnk li hal)* are set aside for next year's seed, as well as for prime sale. Good but small ears *(cocˀ li hal)* are placed in order separately, but against the same wall as *nīnk li hal.* The technique is to make one layer of ears, with the husks left on, tightly packed and all pointing in the same direction. In the next layer, the ears are packed over the interstices of the lower layer, and faced in the opposite direction. As was pointed out in the discussion on insect pests, unless the husks are moist when picked, water is sprinkled over each layer as it is formed, and the layer is then often lightly dusted with Gamaxan insecticide. The end product is a solid wall of maize so tightly packed that rats find it difficult to penetrate, and so thoroughly impregnated with insecticide that damage by weevils is kept to a minimum.[37]

Xsaʔ li hal fills household needs for about a month, since not only small, worm-eaten ears are included, but also ears which have swollen so large as to break open their husks. After this supply has been exhausted, milperos must turn to their store of good maize. From an average *manzana* of milpa, a man can expect about 7 *quintales* of *xsaʔ li hal* or poor ears, 7 *quintales* of *cocʔ li hal* or medium quality ears, and 32 *quintales* of top grade ears *(nīnk li hal)*. When carrying these products home from the granary, the milpero tries to bring enough to supply household needs for a week. For a family of five with an average number of hogs and fowl, these needs would amount to approximately 2 *quintales* of cob corn.

Since carrying from granary to hamlet continues all year long, this is one of the most time-consuming of all agricultural activities. Whereas ten sample milperos expended a mean of only 53 man-hours per *manzana* storing, they spent 109 in carrying. The carrying activities must be effected by the milpero in whose granary the maize is stored. The single exception to this rule appears to be the case when, in time of scarcity just before subsequent harvest, a milpero with more than enough maize stored for his family's needs will lend, without interest, limited amounts of grain to a relative or close friend. Repayment is expected as soon as the new crop matures.

37. The storage methods employed in Chichipate seem to be widely dispersed throughout Maya territory. Reina has presented a very similar description for San José: "A wooden platform *(tapanco)* is constructed, on which the maize is stored. There are two ways of storing cobs: upright, with the broad end of the cob on the bottom for the first layer and in reverse position for the second, or in horizontal rows with the pointed ends of the cobs meshing. The milperos prefer to store their corn in the fields rather than in town, where air currents do not circulate as freely and domestic animals have access to maize. . . . During the dry season, milperos must spray water on the corn to 'raise the humidity level.' This is also done to protect the corn from weevils. In addition, lime dust is spread over the pile and in between the layers of corn to protect it from rats. 'Lime in the eyes of rats leaves them blind'" (Reina, 1967: 16).

Shelling (tenoc ixim—*literally "beating maize"*).—Maize which is sold must usually be shelled first. In contrast to shelling for household use, shelling for sale is usually done in the granary. Often a young boy is hired to do the job, or at least to help with it. The daily pay is the same as that for other agricultural labor: 50 cents in Guatemalan currency or 100 ears of maize.

Though ears of maize are referred to as *hal*, shelled maize is known as *ixim*. The shelling technique is simple but effective. Ears are first husked with the aid of a wooden husking pin. They are then placed loosely in a gunny sack. The worker takes a pole *(che?)* some 38 inches long and 1½-2 inches in diameter and beats the sack until most of the kernels have fallen from the ears. He opens the sack, removes the cobs, and pours the kernels into another sack. Any residue on the cobs he removes by grabbing the cob in his left hand (if he is right-handed), and pressing down on the kernels either with his right thumb or with a shelled cob held in the right hand.[38]

It is obvious that, in order for this shelling technique to be practicable, the kernels should be fairly loose, and therefore the ear must be bone dry. Because of this fact, large-scale shelling is usually postponed until after dry season milpa has been sown, i.e., until after November 15. Labor requirements for the procedure are modest. The mean for a sample of ten milperos was sixteen man-hours per *manzana* in 1965. Two milperos had bought simple, hand-operated shellers, and their work was the least onerous of all.

Maize for home consumption, shelled at home in small quantities at a time, is prepared by hand and often by the woman of the house. After shelling the cob, she lets the grain soak overnight in water and grinds it together with a little lime on her metate. This lime is purchased in a nearby hamlet called Cahaboncito, and sells for five cents a pound. Households use an average of about one pound per week. Between shelling, soaking, grinding, shaping, and baking, the manufacture of tortillas occupies more of a woman's time than any other single activity. To prepare breakfast, she must rise long before daybreak.

Shelled maize is kept in coffers and large clay pots beside the hearth. To discourage rats from lifting up the lids of these bins at night, heavy rocks are set on top of them. A household without at least a few days' supply in these storage bins is practically incon-

38. For comparable shelling techniques, cf. Steggerda, 1941: 109, and Redfield and Rojas, 1962: 45.

ceivable. As soon as the supply runs low, the milpero is sent to bring more ears from the granary.

Related activities.—Immediately after the end of harvest comes the Feast of Saint George, November 15. The image of Saint George was bought by one of the more successful milperos some four years ago, at a time when he thought that the community needed the protection of a second saint. After keeping the image in his home for some time, he sold it outright to the community, and it was placed on the chapel altar. There it remains, except for the times when it is feted in the days around November 15.

A number of scholars have argued that the fiesta system, with its related *cargos,* has served and continues to serve as the major cohesive link between urban centers and peasant communities in the Maya area (Vogt, 1961; Coe, 1965; Haviland, 1966). This *cargo* system has survived in Chichipate through the annual appointment of four *chinamas* for each major saint's day. These four men not only arrange for the image to be carried to one of their houses, where it is ritually fed, but they also pool their resources to buy *aguardiente* for the drinking party which is held on the evening and night following the ceremony. This liquor is sold at the foot of the *ermita* altar, and proceeds from its sale go into a fund with which more liquor may be purchased the following year. The sponsors also arrange for a group of marimba players to provide music for the occasion, and they distribute food to all present in the fiesta.

This pattern of conspicuous giving, so familiar in peasant communities throughout Latin America, has been voluntarily introduced into the frontier community of Chichipate under no direct pressure from anyone outside. Priests, landowners, and other representatives of the larger society are conspicuously absent from the ritual picture. The saint's day fiesta, then, must respond to certain needs that are spontaneously felt on the part of the Kekchi, and not merely reflect the socioeconomic pressure which a dominant class imposes on unwilling peasants, as Marvin Harris has suggested is the case throughout Ibero America (Harris, 1964: 25-43). Chichipateños could have completely neglected the onerous duties associated with fiesta sponsorship in their new setting, and no one would have been the wiser. The fact that they have insisted on fiestas and fiesta sponsorship speaks for the vitality of the tradition in indigenous culture, even when Mestizo pressure is absent.

The manner of celebrating Saint George's Day derives from the prototype not only for other such celebrations, but for practically all ritual performed in the community. Before the saint may be removed from the chapel altar, he must first be well fumigated with copal and placated by prayers, liturgical and extemporaneous, as well as by burning candles. For his journey from the chapel to the home of the fiesta sponsor, he is carried on a litter borne by the sponsors and protected from the elements by a crude, homemade canopy. Once in the home of the major fiesta sponsor, he is placed on the household shrine, is offered more candles, copal, and prayers, and then is fed *atole,* chicken or turkey broth, pork, tortillas, tamales, and bitter chocolate, as well as occasional exotic foods such as watermelon, calabash, and oranges. The floor is covered with pine, and flowers are brought to further adorn the altar. Until liturgical and extemporaneous prayers have been completed and these offerings made, none of the guests may be fed. The foods distributed to these people are of the same type as those which have been offered to the saint. By the time all guests have finished their banquet and listened to the marimbas, the foods offered to the saint have cooled, and they are then distributed to the children present. As soon as the children have eaten their fill, braziers of copal are again lighted, the saint fumigated, and the procession returns with the saint to the *ermita,* where cases of liquor are broken open, the marimba band picks up its beat, and dancing and drinking continue throughout the night, or at least until the liquor gives out.

Such a heavily Christianized rite can hardly be called a Kekchi harvest celebration. Indeed, Chichipateños claim that the only reason Saint George was chosen over another saint was that the man who bought him wanted a saint not found in any nearby community. When he ordered the image from Guatemala City, however, and especially when the village made the subsequent decision to adopt Saint George as the more important of its two patron saints, the fact that this saint's day falls just after the wet season harvest has finished may hardly have been a mere coincidence. Certainly today the ebullience that normally accompanies the completion of harvest finds its institutionalized outlet in this celebration.

The other saint's day, that of Saint Dominic, falls in the lull at the beginning of August when milperos have finished planting and weeding, and before they can begin harvest. The image of Dominic was purchased originally by milperos living south of the Cobán-El Estor road, while that of Saint George was purchased by those living north of it. With the final dissolution of the Tinajas cooperative and the sale of the southern half of Chichipate to a Guatemalan absentee landlord and the northern half to the International Nickel Company, a strong trend has developed for the community itself to divide into opposing moieties. One symptom of this trend is the fact that today sponsors and guardians for the Saint Dominic fiesta all live to the south of the road, while those for the Saint George fiesta all live to the north. Given continuing community-wide involvement in festive dancing and drinking for both celebrations, however, it

Feast of Saint Dominic.

seems significant that on St. George's day, coming at the end of harvest, the most food and liquor are available, the celebration lasts longest, and only then do masked dancers appear.

The most frequently presented masked dance is one which buffoons the Spaniards, the dance of the Conquistadores. The survival of this dance form is particularly interesting, for harassment by Spaniards and Mestizos, so common in the highlands, is much diminished in the frontier situation. The overriding fact that confronts us then is that, left to their own devices, these Kekchi-speaking people have done their best to maintain intact fiesta traditions characteristic of a long subjugated highland past.

Dry Season Milpas (sak?icuaj)

Every milpero in Chichipate planted a dry season milpa in 1965. The degree of success they had, however, varied enormously with the site they selected and the care they gave to the enterprise. Some harvested practically nothing for their efforts.

Because of greater unpredictability of harvest size for dry season milpas, these are seldom as extensive as are wet season or "fire" milpas. In general they range from 5 *tareas* up, the largest not surpassing 3 *manzanas*. Low-lying, well-watered land is the type most frequently chosen; in fact, Chichipateños claim that it is impossible to successfully grow dry season maize on sloping land. To strengthen their argument, they point out that dry season milpas are practically nonexistent in the craggy lands of Alta Verapaz.[39]

As the reader already knows, before the sale and subsequent division of Chichipate's lands, most milperos cropped on the higher plain north of the road in the wet season and on the lower one running between the road and the Polochic River in the dry season. This highly satisfactory arrangement was abrogated with the division of the community's lands, and milperos in 1965 were engaged in risky experimentation as they searched for sites on either one or the other side of the road that would yield harvests in both the wet and dry seasons. The facts that lands well enough watered to produce good crops during the dry season tend to flood in the rainy season, and that the floral components on dry lands tend to differ considerably from those on wet lands, indicate how slim are the chances that such a search will meet with success. The milperos' experimentation, of course, is not blind. In choosing wet season milpa sites for dry season farming, they concentrate on those plots which already have given good yields. Should the milpero conclude that none of his recently harvested wet season milpas look promising for dry season extension, he reverts to using low fallow of one to two years' growth. In 1965, only about half of the dry season milpas were in places where wet season milpa had just been harvested. Most of the remaining were either in places where dry season rather than wet season cultivation was regularly practiced, or where depressions and/or well-watered flatlands were covered with low fallow. Two milperos had become so discouraged over the prospect of finding good dry season crop land on their side of Chichipate that they traveled miles away to the banks of the Polochic, near the confluence of that river with Lake Izabal. There, on sites they be-

39. This accords with the information Drucker and Heizer supply for La Venta. There, steep slopes are rarely planted to dry season milpa, for the soil dries out too much and the maize generally withers (Drucker and Heizer, 1960: 40).

lieved to be part of the national reserve, these milperos received yields of more than forty *quintales* of shelled maize per *manzana*—a return viewed as incredible by their fellows in Chichipate.

Should the dry season crop look promising, the field is often planted to velvet bean *(quēnkʾ caballo)* long before the maize matures. In this manner, the plot may be used year after year, and the need the milpero has for expanding his clearings may be reduced.

Site selection for dry season milpa is accompanied by ritual like that used for selecting the wet season milpa site. Before initiating slashing, the same type of ceremony is also conducted as that which was carried out when slashing began for wet season milpa.

Dry season milpa size depends not only on the availability of suitable land, but also on the price which the wet season harvest is bringing. If that price is good (e.g., around six quetzales a *quintal*), milperos crop as much dry season land as possible.

Planting time is usually October or November on drier land, and December on humid land. In some years, however, the weather changes brusquely about this time. The nights turn cold, the sun scorches the earth, and no rain falls. When this happens, a second planting may have to be made about the fifteenth or twentieth of December. Ideally, the first half of the month should have steadily decreasing rains, and it is under these conditions that dry season milpas do best.

Clearing begins around October 15 and can continue up to January 15, with respective harvests falling from February 15 to June 1. In places where wet season harvest has just been completed, the site is prepared by finely chopping all debris with a machete and spreading it evenly over the ground.[40] Around stumps and trunks that still remain from the clearing which was made at the end of the previous dry season, herbaceous growth is uprooted by hand. If these trunks are piled together in heaps, these heaps are scattered. Rats, much more bothersome in the dry season than in the wet, nest wherever they find protection. By scattering the logs, the milpero hopes to discourage their activity. A good layer of leveled mulch for dry season milpa should range from 5 to 8 inches in thickness.

40. Drucker and Heizer have reported that on La Venta island preparation for dry season milpa is similar. Whereas the debris from the felled bush is burned for wet season milpa, for dry season planting it is merely chopped fine and left as a mulch to conserve soil moisture (Drucker and Heizer, 1960: 39).

Where a site covered by low fallow growth is selected for dry season milpa, after slashing has been completed, some burning often is required. In slashing, the herbaceous growth is the first to go; then saplings are felled with the machete. The larger ligneous growth is simply trimmed, the aim being to eliminate shade. Resulting debris must dry for about a week in the sun before it is burnable. When it is, the milpero goes into his field and burns it by small pieces. If certain areas do not burn well, the debris is spread out and left to dry a few days, when secondary burning is carried out. Generally two burns are enough to entirely clear the milpa.

Milperos prefer, however, not to have to burn over land for dry season farming. The burn leaves only a thin layer of ash, instead of a deep mulch, and so exposes developing plants to desiccation. Milperos do feel, however, that a burned over dry season milpa suffers less from pest infestation than does one where a deep layer of trash has been left on the ground. When the field is heavily mulched, the only defense the milpero has against rodents and other pests is to walk over the trash and frighten them out of hiding. As they try to escape, he and his helpers kill them with a stick or machete.

Prime dry season milpa sites are those which are overrun with velvet bean *(quēnkʾ caballo)*. This legume, relatively new to the area, was first introduced into the Polochic Valley by estate owners. It has rapidly spread throughout the valley, and is today the only crop planted specifically as green manure by peasant milperos. This seems to be an unusual situation for slash and burn agriculturalists working at the technological level found in Chichipate. Tondeur, writing of shifting cultivation in the Belgian Congo, argues that it is unrealistic to expect that green manuring could ever be employed on a wide scale. First, he says, the native grower has no draft animals or machines to facilitate the planting and plowing under of green manure crops. Second, because of the lower return on his labor, he is already obligated to devote nearly all of his time to his crops; it would be impossible to add a technique calling on such a heavy labor expenditure as does green manuring without reducing his crop area and, consequently, taxing his already small resources (Tondeur, 1955: 70). Barrau, assessing the situation in Melanesia, is able to list a few legumes which are grown for their food potential, but he fails to describe any situation in which they are consciously used to improve the soil; in fact, he suggests that the introduction of legumes as a cover crop would be helpful (Barrau, 1958: 52;

Slashing velvet bean for dry season milpa.

1959: 53-55). One of the few reports on shifting cultivation that does mention conscious upgrading of the soil through the planting of specific cultivates comes from Brown and Brookfield's work among the Chimbu. Here they found that casuarina seedlings are planted with vegetable crops on the better soils so that by the time the land is abandoned, the trees are already well established. Apparently the Chimbu know that casuarinas aid restoration of soil fertility; in effect, what these plants do is fix nitrogen through root nodules in addition to providing humus (Brown and Brookfield, 1959: 11, 19).

Velvet bean *(quēnkʾ caballo)* is planted either at the time of the first dry season milpa cleaning or, with wet season milpa, just after maize has broken through the ground. The technique is to dibble a shallow hole and drop seeds three at a time into it, spacing the hillocks some 4 *varas* apart. After two months of growth, velvet bean begins to have drastic effects on the milpa. Where weeds, grasses, and small trees have begun to take hold, it gradually covers them and chokes them out. Within six months, it yields a thick cover of dark green leaves that can reach up to 8 feet in height.

In cleaning milpas covered with velvet bean, workers raise the top of the vine mat and look for the stalks of the bean. These they slash with a machete held in their right hand. The slashing is done as near to the ground as possible. Once the vines have been severed, workers then go back and chop them up very finely. On decaying, the vines leave up to 3 to 4 inches of "manure"—dark, decayed vegetable matter. Though the legumes replenish themselves as the dry season maize matures, they do so gradually, so that by the time they have covered the field again, the maize is either ready for harvest or has been harvested. Milpas planted to velvet bean never revert to grassland, nor will they, if regularly slashed, return to forest. Some plots, therefore, have been used consecutively for fourteen years of dry season farming and show little indication of diminishing fertility.

The present use peasant agriculturalists are making of velvet bean in the Polochic Valley would seem to have major relevance for slash and burn systems in many parts of the humid tropics. It argues that, contrary to Tondeur's expectations, it *is* realistic to expect that green manuring can be employed on a wide scale by people at a relatively low level of technology.[41]

41. The precise benefits of legumes and green manure in improving tropical soils are still in question, however. Urrutia argues that before this question can be settled, basic research must first be done on the processes of nitrification, nitrogen fixation, nodulation, and the soil parameters that affect those processes under tropical conditions (Urrutia, 1967: 17).

Seeding in dry season milpa tends to be a more solitary event than it is with wet season fields. Occasionally work parties are formed, but more often a man works alone or with the members of his immediate household. Spacing between maize hillocks in dry season milpa is the same as in wet season milpa, but both the dibble and its use differ. A pole with at least one knot is cut and sharpened to a very thin point at one end, the tip being no more than 6 inches below the knot. When the dibble is thrust into the ground, the knot prevents it from penetrating too deeply. The technique for handling this dry season dibble is to puncture a small, finger-type hole in the ground and then to retract the dibble immediately, taking care not to move it from right to left, nor from front to back, as is done in dibbling for wet season maize. To give seedlings a chance to breathe, before puncturing the soil, the mulch is removed from a few inches around the spot to be dibbled.

Dry season milpas require only one cleaning, some four to six weeks after seeding, and this is done exclusively with the machete. Since pests of all sorts are more numerous than during the wet season, growing ritual becomes especially important. If the milpero has not presented *huacales* of worms to the household saint during the wet season, he is sure to do so now. And he cannot neglect the ritual offerings that accompany the emergence of tassels on his maturing crop, nor those which customarily accompany the gathering in of first fruits.

Since dry season milpas tend to be small and scattered, harvest groups are also small, and usually limited to the householder and his sons. Yielding at a period of the year when highland stores of maize are seriously depleted, dry season grain usually brings a premium price. The milpero sells as much of it as he possibly can, often indeed leaving his own stores so short that he must later buy or borrow grain to carry him through to the main, wet season harvest.

Dry season intercropping.—The favorite time for starting fields of chili peppers is the beginning of November, just after wet season maize has been gathered in. Women grind dry peppers on their metates and then winnow off the chaff. In the recently harvested fields, milperos pull up all weedy growth by hand, leaving only dried maize stalks. In planting chili, the seeds are merely cast on the surface of this denuded soil and ground in with the milpero's foot. Ideally there should be about 12 inches between plants. A *tarea* of chili consists of 35 *varas* square. From each of these *tareas,* milperos expect an average crop of around 5 *arrobas*. Whether they get this or not, however, and when they get it, depends on whether they intercrop the chili or sow the entire milpa to it. Two fields of chili peppers observed in 1965 were sown at the same time in *melb ru li chʔochʔ*. With the first, maize was dibbled over the chili five days after it had been planted; with the second, the chili was left by itself. By August the intercropped chili was only one-third as large as the other and was just developing flowers, while the other was giving ripe fruit.

Since chili peppers are so essential in seasoning Kekchi food, practically every milpero plants at least 1 *tarea* of them at the beginning of the dry season.[42] Other crops which may be added at this time but that, in reality, seldom are include garlic, onion, watermelon, sugarcane, plantain, banana, manioc, malanga, bush bean, squash, tomato, and sweet potato. For manioc, malanga, plantain, and banana, milperos prefer to obtain "seed" from recently burned land. If, for example, a person finds that the low fallow he plans to burn for dry season milpa contains specimens of these four plants, he will often make no effort to protect them. Rather, he will purposely direct the fire over them, for he considers the sprouts that emerge afterwards to be the finest "seeding" material.

Flooding of dry season sites, once the rains have returned, somewhat restricts their use for intercropping. While sugarcane, plantain, and banana can stand waterlogged conditions part of the year, malanga, manioc, chili pepper, and sweet potato cannot and will not survive under such conditions.

Gleaning and cleaning.—Though, through careless harvesting, some quantities of grain may remain in the milpa, no gleaning is practiced. The most that a milpero will do in a harvested and abandoned wet season milpa is to clear away debris and growth around secondary crops or intercrops. The first of these cleanings is started from one to two months after harvest, i.e., once major dry season planting has been completed. With manioc, pineapple, tobacco, sugarcane, malanga, plantain, and banana, approximately 1 *vara* around each plant is thoroughly cleared away. The trash is chopped and leveled to serve as a mulch. Peanuts, if they are still bearing fruit, are weeded by hand, and then a swath about 1 *vara*

42. Wild chili *(chiltep)* is also collected. (Cf. Kelly and Palerm, 1950: 138; and Wisdom, 1940: 55, for references to such collection.)

in breadth is cleared around them. Other low, herbaceous growth is cleaned in like fashion. Only sweet potatoes are completely left to fend for themselves.

Cleaning of this type requires only three to four man-days of labor per *manzana* of *rok cuaj* (harvested milpa). Seldom is any attempt at cleaning made unless competing growth has reached more than 1½ meters in height. Any given milpa may be cleaned, then, only once every four to six months. To the untrained eye, a first-year extension milpa is indistinguishable from low, secondary growth. But just as in many other tropical, intercropped systems, this very untidy field continues to provide important supplements to the basic staples of the diet.

Wet Season Recultivation

Individuals hard pressed for wet season milpa sites may be forced to revert to those they had used the previous year. Seldom do they do this, however, if wet season cultivation of the site was followed immediately by dry season cultivation. Nor can they do it more than one time, for even at the first try, the yield tends to be low.[43] The most common circumstance under which a milpero will most likely thus press the resources of the land is when, due to an incomplete burn or unusual flooding conditions, he has received very little from his wet season efforts the previous year. In this case, he feels he must try to rescue at least some of the arduous labor involved in clearing the land the first season. Otherwise, he goes back over the same wet season site in successive years only when nothing else is available. Because of the scarcity of well-drained land on the south side of Chichipate, this "last recourse" was unusually common in 1965. To the north side of the road, however, where drainage is much less of a problem, no one was engaging in wet season recultivation.

One dilemma which wet season recultivation presents is that intercropped materials are disturbed at the height of their bearing capacity. If a milpero has these plants (e.g., plantains and manioc) in sites other than the one he is recultivating, he simply lops off some of their leaves as he clears the land for recultivation; in most cases the plant recovers fairly rapidly. Should these be the milpero's

43. The major exception to this generalization occurs with lands that have been cleared from mature forest. Here, because of competition from dying forest root systems, the first year of cultivation may yield a lower harvest than does the second year.

principal intercropped items, however, he makes elaborate protective clearings around them.

Succession to Grassland

Interesting in the light of various theories that have been proposed for the collapse of Classic Mayan culture in the Petén is the problem of grass encroachment. Such encroachment, if it were to become widespread, would be disastrous for a system of cultivation like that practiced in Chichipate. Grass not only competes with cultivated plants for available nutrients and space; it also harbors rodents. From his work in the Polochic Valley, Popenoe concluded that during the early years of grassland, the rodent population undergoes a very accelerated growth. In his words: "When grass is plentiful in cornfields, this effect is reflected in the large number of young corn plants either killed or injured from physical damage inflicted by rats and mice. After the grassland has been established for a few years, the number of rodents builds up rapidly to a peak and then declines as natural enemies appear on the scene" (Popenoe, 1960: 61-62).

In the system of cultivation practiced in Chichipate today, no regular use is made of grasslands, either for animal fodder, rope making, or thatching. Chichipateños look upon grass entirely as a liability. Dibbling is ineffective in grassy plots; even if maize does get started there, it is quickly choked out. Burning is equally ineffective.

Fortunately for the Chichipateños, only small marginal plots had become overrun with grass by 1965. The whole region had been so recently carved out of mature forest that apparently there had not been time for anything more extensive to develop. Even the stoloniferous *pachʔayaʔ* (*Paspalum conjugatum* Bergius) lasts only a maximum of three years in the frontier setting. By that time it has been completely killed out by developing sapling growth. Should the milpero wish to rid a plot of the pest sooner, he always has recourse to *quēnkʔ caballo* (velvet bean).

The five worst grassland associations, from the standpoint of threats to standing crops are: *pachʔayaʔ* (*Paspalum conjugatum* Bergius), *caki pachʔayaʔ* ((*Axonopus compressus* (Swartz) Beauv.)), *qʔuim* (*Paspalum caespitosum* Flugge), *qʔuim* ((*Paspalum botterii* (Fourn.) Chase)), and *qʔuim* (*Scleria pterota* Presl.). Other common grasses and sedges have already been enumerated on pages 93-94. They are most likely to become established in milpas which are poorly drained and which are marginal for wet season cultivation anyway. Otherwise, grass appears only near paths, where seeds have been scattered by workers as they walk to their fields, or in those isolated spots where a given plot has been burned over for a second consecutive year. Here grass appears mainly because the second firing, while destroying young tree species, does not seriously affect solid mats of grass roots that lie under the ground.

It is in dealing with grasses and sedges that milperos use their hoe, or *asaron*. Stoloniferous grasses that have developed a heavy mat of roots are very difficult to kill. For one's effort to be effective, the root must be dug up, dried, and then burned, a very time-consuming task. In light of the still relative abundance of land in the Chichipate area, it is no wonder that milperos prefer to abandon a piece of land rather than laboriously clear it of grasses.

Peasants in Chichipate have been told by some outsiders that the use of plow culture would obviate the problem of gradual succession to grassland. The experience of the owner of a large, mechanized *finca,* a few miles up the road from Chichipate, has led them to question this suggestion, however. There, a tractor and bulldozer were used to clear over the land. Then the land was cropped on a year-round basis, alternating between dry season maize with inter-cropped beans, wet season maize, and at times rice, for a continuous period of five years. By the end of that time, the entire area had become so overrun with grass that, even with the use of a tractor and plow, the land could be utilized for nothing other than pasture. Milperos in Chichipate, having observed the course of events, declare that the only reason succession to grass has not yet become a major problem in their fields is that, even in those cases where they use a given plot for two successive years, they burn it rather than turn it over with a plow. The experience has served to reinforce their belief in the superiority of slash and burn techniques.

Recent work on Maya prehistoric sites indicates that Chichipateños may be right when they assert that the problem of grass encroachment can be controlled with a properly handled slash and burn system. Pollen studies made by Cowgill and Hutchinson suggest that grasses were more prevalent prior to extensive agriculture in the Petén than they have been since (Cowgill and Hutchinson, 1963a). This seems in accord with what Hester wrote over ten years ago, namely, that there are no perpetual grasslands in the lowland Maya region occupying soils which otherwise would be well suited to maize agriculture (Hester, 1954: 41).

In 1965 there were no cattle or horses in Chichipate.[44] Should the time ever come when they begin to keep these animals, the view of the milperos toward grass will probably change significantly. For now, however, they seem to be successfully coping with the grassland problem. *Quēnk' caballo* constitutes their major weapon. Its use seems of primary importance in evaluating the potential of slash and burn systems for lowland Guatemala.

Fallowing

In all tropical slash and burn systems, the length of fallow should surpass that of any of the cropping stages. Unless this is so, cropping sites customarily suffer from ever increasing weed encroachment. This threat is overtly recognized by milperos in Chichipate, who insist that it is more weed encroachment than loss of fertility that leads them to abandon a piece of land. Kelly and Palerm (1950: 100), Morley (1947: 135-36), Heizer (1960: 217), and Stevens (1964: 304) would all agree with them.

This is not to say that there is never fertility loss when a milpa is repeatedly used. In a slash and burn system, cropped specimens take nourishment more from the concentrated organic matter of the ash and decaying vegetation than from the soil itself. The purpose of burning is to convert the energy stored up by the vegetative cover into a more usable form. But when that cover is thin, by the time most of the available energy has been converted into smoke and heat, very little is left in ash.[45]

44. This statement may seem odd in light of the fact that Chichipateños have built a few crude cattle-type fences, have strung barbed wire, and serve beef to helpers at seed time. Those cattle fences which do exist, however, have been erected either to protect milpas from animals that occasionally roam in from nearby estates, or in anticipation of the grazing complex the owner of Yuscarán hopes to establish on his portion of Chichipate lands. The beef which they occasionally consume, and which they distribute at planting time, is obtained from traveling merchants who come in regularly on trucks and buses.

45. The experience of Chichipateños and others runs counter to that recorded by Ursula Cowgill. On the basis of 152 soil samples, she has concluded that there is a significant decrease in all nutrients and pH with repeated use. Second year plots she has studied show more deficiency symptoms than first year fields. At the same time, she has claimed to have found no significant difference in the number of weeds per acre between first and second year fields. She thus concludes that the observed yield decline is due to a loss in soil fertility and not to weed competition (U. Cowgill, 1960: 1010).

The length of fallowing practiced by the milperos of Chichipate is difficult to determine. First of all, no written records are kept. Second, seldom does a milpero keep a given milpa to any exact size. Though milpas tend to be roughly square or rectangular, they are regularly expanded along their borders, so that they seldom are the same size the second time they are used as they were the first. Third, secondary growth develops at a differential rate in each fallow, depending on such variables as the lay of the land, texture, organic and chemical content of the soil, surface water, and surrounding vegetation. After the first year of fallow, even expert ecologists are hard put to determine exactly the age of the vegetation. In good, well-drained locations on the shores of Lake Izabal, they sometimes cannot distinguish late secondary growth from early climax forest. A fourth difficulty in obtaining exact figures on the length of fallow derives from the practice of continuing to clean and harvest root and tree crops long after a site has been withdrawn from the maize-cropping cycle. Though, in most cases, this extension milpa is actively exploited only during the first year or two after the final maize harvest, there are isolated cases where banana, pineapple, or sugarcane are maintained for most, or indeed all, of the fallow period.

In spite of repeated declarations to the effect that climax growth is that most preferred for preparing wet season milpa, there was no indication in 1965 that milperos were allowing their fallows to reach this stage of development. Their verbal judgment was that five to six years of fallow would be ideal. In practice, however, it seems that they were settling for much less.

As they select milpa sites, milperos try to pick spots in the general vicinity of previous years' milpas. They tend to expand the size of a given site the second time they use it. This is easy because surrounding vegetation often dies due to either the intensity of the original milpa fire or destruction of unprotected surrounding timber by high winds. The following two individual case-diagrams illustrate both the method of expanding one's milpa and the time interval that is becoming regularized in Chichipate, as of 1965.

The most striking insight to emerge from these rough sketches is the variability of the fallow period, even when the same milpero is involved. Land which gives good yield is apparently used repeatedly, until it shows signs of seriously diminishing. Thus, plot number 4 in the second diagram was used three times, at only two-year intervals. Note also the frequency with which a given

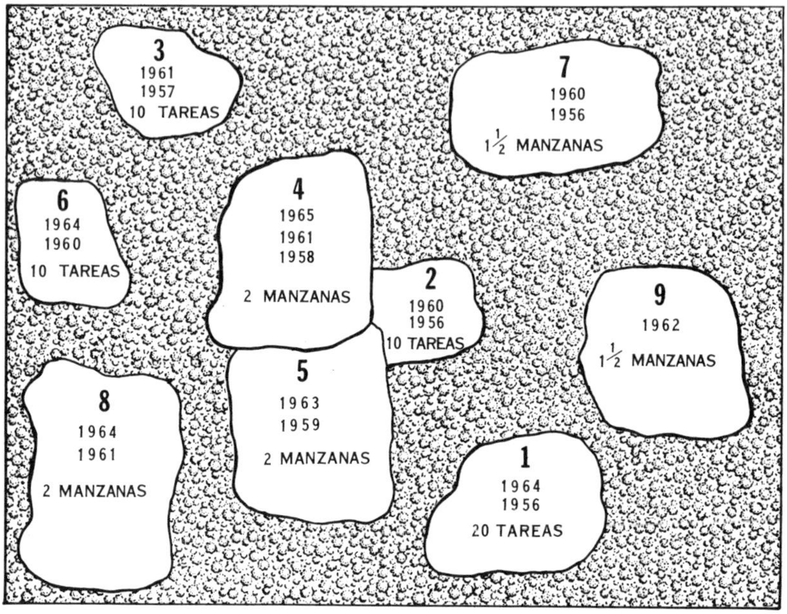

The method of milpa expansion and the time interval that is becoming regularized in Chichipate.

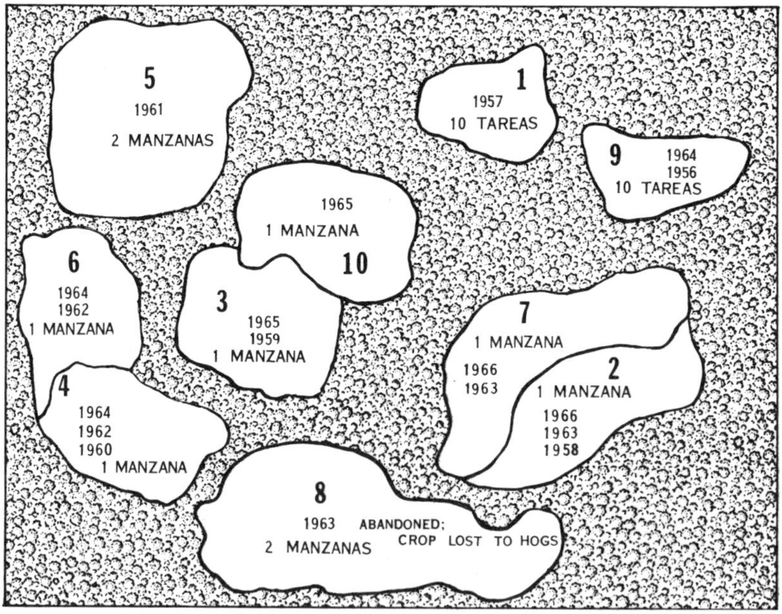

milpa is opened up the same year the milpero has returned to adjoining land, as well as the increasing size of the plantings throughout the years. Of all the milpas, plot number 1 of the second diagram has been left fallow the longest period, eight years. The average time span between the first and second wet season milpa has been only 3.4 years, a questionable period from what is usually said about tropical ecology, but well in line with some of the other reports for the lowland Maya area. Ursula Cowgill found that, of those informants who planted a single crop before moving to another location, three-fifths asserted that only two years of rest would be needed before the land could be used again, another fifth favored a three-year rest, and the remainder gave replies of one to five years (U. Cowgill, 1962: 276). Drucker and Heizer, on La Venta Island, found that tracts were permitted to lie fallow for a period ranging from three to five years, with four years being regarded as optimum (Drucker and Heizer, 1960: 40). In his report on the Polochic Valley, Popenoe states that the fallow period there lasts from three to four years, but observes that as one proceeds northward through the Petén and Yucatán, the length of fallow increases as the country becomes increasingly drier (Popenoe, 1960: 64). Given the problems of weed encroachment and fertility loss, one may ask why milperos reuse plots after such a short fallow period. The answer for Chichipate seems to be threefold. First, yields received from recently opened lands tend to be higher than those obtained from the exhausted soils of the highlands. Second, they have not yet built up adequate resources of fallow land. Third, and perhaps most important, the absentee landowners of Chichipate have forbidden further encroachment on mature forest growth. This proscription has brought some milperos to a desperate state. One, in 1965, said that he would have to find some mature forest growth for milpa the following year in order merely for his family to survive. "I am going to die," he bemoaned, "if there is no place to do it." The solution many of his companions keep in the back of their minds, of course, is possible migration farther back into the forest, perhaps into the Petén itself.

Except for the claims of absentee owners, land is theoretically a free good among the Kekchi. As has already been pointed out, however, in practical fact they maintain usufruct rights to it as long as they wish once they have first cleared it. Ideally, each milpero attempts to build up a reserve of secondary growth that will supply him indefinitely. Thus, should he plan to return to the same plot once every seven years, he would need 7 *manzanas,* or 175 *tareas,* of land for every *manzana* he wished to cultivate. A sizeable number of the older residents of Chichipate have been able to establish just such reserves, though the division of the community into two separate property units has cut into some of their rights. None of the households arriving within the past two to three years are in such fortunate circumstances, however. Their foothold in Chichipate is exceedingly precarious. They often are reduced to clearing secondary growth on land without permission, and this rapidly gives rise to intense interpersonal conflicts.

Unlike other slash and burn peoples, the Kekchi of Chichipate seem more oriented to plots than to plants.[46] Though they may have tree and root crops maturing in the fallow, they still speak of it as land at rest *(al c*ʔ*al).* It is impossible to conclude when cropping ends and strict fallowing begins, for the two occur simultaneously. Dry season milpas, furthermore, tend to be planted in patches within the perimeters of wet season clearings. The result is that any given piece of fallow may represent a whole series of different stages on the return to secondary forest growth. When the milpero states that he is going to return to a plot he has not used in five years, he generally refers only to its previous use for either wet season or consecutive dry season maize. Root or tree crops are only incidental to fallowing, as his expression "I have a few plants in my fallow" *("cuan li acuĩmk sa*ʔ *lin rok cuaj")* indicates.

Succession to mature secondary growth.—Chichipateños recognize eight major normal stages of development in the replenishment of forests, based solely on the variables of thickness of brush and size of ligneous growth:

Rok cuaj	A field covered with small plants which appear in the first few months after harvest (literally, dried corn stalks).
*Coc*ʔ *pim*	Low weeds.
Pim ru	Thick, low weeds.
Mas pim ru	Very thick brush.
*Coc*ʔ *che*ʔ *ru*	Sapling stage of secondary growth.
*Nĩnki al c*ʔ*al*	High secondary growth.
*Mac*ʔ*a*ʔ *xpimal*	Higher secondary growth, characterized by sparse undergrowth.
*Nĩnk li q*ʔ*uiche*ʔ	Mature forest growth.

46. Cf. Conklin, 1957: 134 for an example: "To re-emphasize an important point: Hanunóo agriculture focuses on plants not plots."

There are no clear cut-off marks between one and the other of these vegetation complexes, though once a given complex is well established, there is little dispute as to the category it best fits. Nor, as has already been indicated, is there any precise time requirement for the development of each successive stage.

In terms of vegetative type, most wet season milpas were being cleared in *mas pim ru, coc? che? ru,* or *mac?a? xpimal* in 1965. Some milperos, in fact, were declaring that, even if unlimited virgin lands were to become available, they would not let their fallow revert to *nĩnk li q?uiche?*, for clearing such land would entail as much labor as they expended when they first arrived in the hamlet. They do, nevertheless, prefer that numerous ligneous type plants characteristic of mature forest be present before removing vegetation for repeated cultivation. In this way, they not only avoid the worst of herbaceous weed infestation, but benefit from the nutrients present in a heavier ash layer. Those ligneous species they most commonly look for are:[47]

Kekchi	Binomial
Hu	*Virola guatemalensis* (Hemsl.) Warb.
Guamo	*Tabebuia pentaphylla* (L.) Hemsl.
K?eki che?	*Swartzia guatemalensis* (Donn. Smith) Pittier
Sal tul	*Pouteria mammosa* (L.) Conquist
Cuãchil	*Dialium guianense* (Aubl.) Sandw.
Puj	*Ochroma lagopus* Swartz
Inup	*Ceiba pentandra* (L.) Gaertn.
Ch?abay	*Guazuma ulmifolia* Lam.
Balamax	*Luehea* sp.
Mocõch	*Orbignya cohune* (Mart.) Dahlgren ex Standl.
K?ante?	*Gliricidia sepium* (Jacq.) Steud.
Chi?	*Byrsonima crassifolia* (L.) HBK
Camãnc	*Neurolaena lobata* (L.) R. Br.
K?inam	*Vatairea lundellii* (Standl.) Killip ex Record
Obel	*Piper umbellatum* L.
K?erk	*Heliconia latispatha* Benth.
Sutz?ujl	*Swietenia macrophylla* King

As fallow plots gradually fill in with these ligneous plant types,

47. The reader may wish to compare this listing of common species in medium secondary growth, as perceived by Kekchi informants, with a much more complete one made for the southern Petén by a professional botanist (cf. Lundell, 1937: 194).

they become the source of material for house construction and firewood, and also provide occasional fruits, medicines, and useful saps. This adds emphasis to the point already discussed: that the fallow plot is far from a useless commodity. Still, the major explicit purposes of fallowing are to combat weeds and to restore fertility to a given plot of land, and these purposes are interpreted specifically in terms of maize-producing capacity. Except for those cases where velvet bean has been allowed to take over a plot, milperos realize that fallowing is an obligatory requisite for continued production of the staff of life.

Shifting cultivation.—Fallowing, as practiced by Chichipateños, necessitates regular shifting of cultivated sites, though not of the hamlet itself. For such a situation, Allan, in his *African Husbandman,* has suggested that the term "land rotation cultivation" be used instead of "shifting cultivation" (Allan, 1965: 6).

According to Allan, the term "shifting cultivation" may be applied only to those situations in which the whole community of cultivators move. Such a move may be broken down into two basic types, "obligatory" and "voluntary." Whether "shifting cultivation" is "obligatory" or "voluntary" seems to depend solely on factors of natural environment and demography. One of Allan's examples of "obligatory" shifting cultivation is a situation in which land largely unsuitable for cultivation is exploited. A periodic move from one habitation site to another thus becomes a matter of agricultural necessity. In contrast, he feels that "voluntary shifting cultivation" may be found wherever "land is so abundant, in relation to the population and its requirements, that the period of natural soil regeneration has no practical significance, and a man need not think in terms of returning to land he has cultivated within the foreseeable future. . ." (Allan, 1965: 6).

Throughout this monograph, the term "shifting cultivation" has been used in the sense that Allan uses the term "land rotation cultivation." The system employed by Chichipateños will become "shifting" in Allan's sense only if, at some future date, most milperos abandon their hamlet. Such a move will probably lie somewhere betwen Allan's "voluntary" and "obligatory" poles, though the factors involved will probably be much more cultural than demographic and environmental. Present land resources seem adequate for a permanent settlement, given the current population and level of technology. With the squeeze that is resulting from the combination of population increase and ever more stringent land use restrictions, however, this situation may not hold true for long. The conflict of two cultural systems, the one governing labor and the other governing land, may eventually bring collapse to the agricultural system as it exists at present.

Section III.

Shifting Cultivation and Guatemalan Development

Calculating the potential of a shifting system of cultivation is always a difficult task. Fields lie scattered, are irregularly shaped, and vary in size from season to season. To further complicate things, intercropped items are often planted and harvested a little at a time, so that not even the grower has a clear idea as to their yield.

Lacking an inordinate amount of time and patience in the field, it is unrealistic to hope to obtain exact measures for every single item produced in milpas like those of Chichipate. However, with maize, the main starch staple, the situation is not hopeless. It would be possible to measure the exact amount of stored grain in a large sample of recently filled granaries. Correlating this with the size of the harvested milpas would give a fairly accurate picture of yield.

Unfortunately, I was never able to be in Chichipate at the time of final completion of harvest. I was forced, then, to rely upon informants' estimates as to the productivity of their milpas during the current and the previous year and, on this basis, to work out average yield and productivity figures. There is a pressing need for someone to go into the lowland Maya area at the time of harvest and take exact measures. Until that happens, however, rough estimates seem better than nothing, especially when they demonstrate internal consistency and are in general agreement with estimates collected by other investigators.

The figures that follow concerning labor input per *manzana,* yield, size of milpas, average production per household, and patterns of consumption are based on material obtained in interviews with thirty milperos (i.e., one-third of Chichipate's household heads

A nuclear family of Chichipate.

as of 1965). Most of them refer to wet season cropping, since it constitutes the major productive activity of the community.

The first figures which will be presented will be those referring to labor input. Estimates, in this regard, vary with the type of land that is used. They help explain why, though in theory men prefer mature forest in selecting milpa sites, in practice they more frequently turn to low fallow or sapling growth. The clearing of mature forest does have the advantages of yielding a milpa site relatively free of weeds, and of eventually adding to one's total store of fallow land. But it has the disadvantage of being carried out on lands that have not yet met the test of maize production and are therefore something of an unknown quantity. At times these lands turn out to be root-bound and yield a poor harvest the first year. At other times they flood when heavy rains come. Above all, they require a heavier input of labor than do low and medium fallow lands.

The total average labor input for a *manzana* of land cleared from mature forest is reported to be 888 man-hours; for land cleared from sapling growth it is 725, and for land cleared from low fallow it is only 650.[1] The time saved by avoiding weeding is more than lost in the heavy demands of felling. It seems understandable that, in spite of the problem of smaller yields, 68 per cent of Chichipate's wet season milpas were planted in low fallow sites in 1965.[2] Sites covered by sapling and mature growth, it will be recalled, were used for only 21 per cent and 11 per cent, respectively, of the community's milpas. Proscriptions on the part of absentee landowners undoubtedly have their effect on the situation, but they do not seem to be the only factor involved.

Reina (1967: 5) has suggested that with milperos of San José, the time-labor needs are greater during clearing and preparation of the plot than they are after planting. As will be seen, such does not seem to be the case in Chichipate. Nor is it the case in other well-documented studies of slash and burn systems (cf. Conklin, 1957: 150). It should perhaps be noted that Reina gives no detailed analysis of the time-labor needs of the San Joseños; the reader is left uncertain as to the informational basis of his statement.

Table 16 lists averages for most of the labor input required to prepare, maintain, harvest, and dispose of the fruits of wet season milpa. Ritual, preparation and care of implements, and traveling to and from the milpas are all included in the estimated labor demands for each major phase of the milpa cycle. Time lost

1. For similar slash and burn activities in Mexico, Poleman has calculated the use of approximately 570 man-hours per hectare. Though Poleman's estimate is an extremely rough one, it does generally correspond with these figures obtained in Chichipate (Poleman, 1964: 82).

2. The selective use of lands in Chichipate corresponds closely with that reported by Drucker and Heizer for La Venta Island: "The *acagual* is easier to clear, not only because the plant growth is smaller in size, but because the second-growth consists almost entirely of soft woods. It burns off cleaner, making possible evener and more complete planting, as well as easier weeding and harvesting. . . . The modern milpero prefers, and beyond a doubt his pre-Columbian ancestor preferred, to work in *acagual*. He clears *montaña* only when his *acaguales* no longer suffice or when he wants to extend his clearings for pasture" (Drucker and Heizer, 1960: 39).

due to illness is not accounted for, however. This may be considerable, though no specific information regarding it was collected. It is worth remembering that a goodly number of milperos and their wives and children suffer from chronic malaria.

The table indicates no time expended in felling when milperos use land covered by medium and low fallow. This is due to the fact

TABLE 16
ESTIMATED AVERAGE LABOR REQUIREMENT
(WET SEASON MILPA)

	Man-hours per *manzana* of wet season milpa		
Activity	Mature forest (8 + years' growth)	Sapling growth (4-7 years' growth)	Low fallow (1-3 years' growth)
Site selection	10	6	6
Slashing	68	113	84
Felling	300		
Firebreaking	10	26	26
Firing	5	2	2
Planting maize	44	44	44
Fencing	5	5	5
Watching	46	46	46
Weeding	2	56	105
Intercropping	103	88	42
Reaping	99	112	97
Granary building	23	23	23
Carrying	104	129	103
Storing	53	57	52
Shelling	16	18	15
Total man-hours	888	725	650

that in such situations, all small growth, including saplings, is slashed with machetes. Fencing accounts for a very low average labor input, emphasizing again the preference most milperos have for sites far enough away from the hamlet so that it will not be required. One suspects that the activity of animal pests varies with the type of site used. Experiments at Murciélagos indicate that pest activity increases when a milpa is planted in low fallow, or used in successive years. Yet, milperos of Chichipate do not distinguish this variation in the estimates of labor demands.

Weeding is reported as much more arduous with low fallow. Intercropping is less time-consuming there, however, than in sites just cleared from mature forest, probably reflecting the milpero's recognition that mature growth sites have greater future potential than do low fallow sites.

A final point worth making is that carrying from granaries to the hamlet is one of the most time-consuming tasks of the entire milpa cycle. Yet this important activity is frequently overlooked entirely in discussions of lowland Maya cultivation.[3]

How well labor input is compensated cannot be judged without an idea as to yield. Combining estimates made by thirty milperos of production in good, bad, and middling land, there emerges an average of 3,845 pounds of unshelled maize per *manzana*. Using the same 55/100 ratio of shelled to unshelled maize that Ursula Cowgill (1962: 277) and Reina (1967: 14) have employed for the Petén, one obtains the figure of 2,121 pounds of shelled maize per *manzana*. Transposed into the acreage system, this amounts to 1,373 pounds of unshelled maize, or 755 pounds of shelled maize per acre. Such an estimated average, representing both first and second year wet season milpas, is a bit on the high side when compared with yield reports that come from the Petén and Yucatán. Cowgill, for example, gives average unshelled yield estimates for wet season

3. A comparison of labor estimate figures from Chichipate with those collected by other workers in the lowland area may give some basis for assessing the validity of the former.

Considerable discrepancy exists between the Chichipate data and those presented by Drucker and Heizer for La Venta Island, where it is claimed that a man can clear over a hectare of "virgin jungle" in 15 to 20 man-days (Drucker and Heizer, 1960: 39). Transposed into the Chichipate system, this would mean from 137 to 181 man-hours per *manzana,* far less than the number of hours reported in the above table as average. Morley's estimates for clearing low secondary growth in Yucatán accord more closely with those obtained in Chichipate. According to him, about one-fifth of an acre may be cleared in one day (Morley, 1947: 143). This, transposed into the Chichipate system, comes to 112 man-hours per *manzana*. A third report, that by Kelly and Palerm, gives an overall average labor figure for the complete milpa cycle. This comes to 88.7 man-days per hectare (Kelly and Palerm, 1950: 121). Transposing this into the Chichipate system, there also seems to be general agreement: on the basis of eight-hour days, it comes to 798 man-hours of input per *manzana,* not including granary construction.

One of the difficulties with all these comparative labor estimate figures is that they either do not specify the type of vegetation for which they are relevant, or else they do not give comparative data for other types of vegetation. Another is the fact that none of them have been broken down to include such common activities as site selection, firing, carrying, storing, shelling, and so forth.

cropping in the Petén as 1,425 pounds per acre the first year and 1,010 pounds the second (U. Cowgill, 1960: 1010; 1962: 276). Estimates from northern Yucatán run from 930 to 1,400 pounds per acre for first year fields and from 460 to 1,100 pounds per acre for second year fields (Kempton, n.d.; Emerson, n.d.; Steggerda, 1941).[4] The Chichipate estimate does fit loosely into the general picture of yield that is emerging for the lowland Maya area, however. Though it must be considered as nothing more than a good approximation, its coincidence with other reports and its high internal consistency seem to indicate that approximation is fairly near to the fact.

By actual count, the average size of wet season milpa holdings in 1965 was 3.1 *manzanas* per household. Using the Chichipate system of measurement, this comes to 8.71 acres, close to the 9.8-acre average milpa size measured by Steggerda in Yucatán (1941: 112). Taking the Chichipate figures and multiplying them by the average reported yield per *manzana,* it becomes possible to make an estimate of total production per household and for the community as a whole. Thus, the average milpa size of 3.1 *manzanas* multiplied by the average reported yield of 2,121 pounds of shelled maize per *manzana* would equal 6,575 pounds of shelled maize per household. Transposed into the English system, this would be: average milpa size of 8.71 acres, multiplied by 754.9 pounds of shelled maize produced per acre, equals 6,575 pounds of shelled maize per household. However, this figure, 6,575 pounds, does not represent the entire amount of maize available, on the average, to each Chichipate household. To it must be added the dry season production estimate of 1,723 pounds of shelled maize, much smaller than the wet season estimate, but significant nevertheless. When this is done, it appears that, on the average, a household in 1965 produced for use and sale some 8,298 pounds of shelled maize.

Average reported daily consumption of shelled maize, for people of all ages, is 1.4 pounds per person per day. This is considerably less than the 1.7 pounds reported by Cowgill for the Petén (U. Cowgill, 1962: 277), but it falls well within other estimates that have been made ranging from 1.3 to 2 pounds per person per day

4. Particular care should be exercised in comparing Chichipate estimates with data from Yucatán. While the latter are undoubtedly more relevant than data from the Guatemalan highlands, they should be considered at best as suggestive rather than conclusive, so far as the southern Maya lowlands are concerned.

(Benedict and Steggerda, 1937; Morley, 1947; Higbee, 1948; Stadelman, 1940; Kempton, n.d.; Drucker and Heizer, 1960). The fact that it is less than the Cowgill figure may reflect, in part, a greater use of intercropped foods in Chichipate than in the Petén. Unfortunately, since we have so little information on intercropping in the Petén, this hypothesis is impossible to either confirm or refute. Given the figure of 1.4 pounds per person per day, however, and multiplying this by 365 days, one arrives at a global average consumption figure of 511 pounds per person per year.

Chichipate households number, on the average, 4.8 persons. The amount of shelled maize needed to feed a unit of this size comes to 2,452.8 pounds per year (very close to Erasmus' "liberal" estimate of 1,300 kilos per annum per family [Erasmus, 1965: 295]). Subtracting this from the average amount of shelled maize available per household leaves a "surplus" of 5,845 pounds.

It is from this "surplus" that milperos make their sales. As harvest begins, truckers come into the community and bid for grain. On the average, 2,083 pounds of shelled maize were sold per household in 1965. As supplies ran short just before the following harvests, however, milperos were forced to purchase an average of 428 pounds per household, leaving a net difference per household of 1,655 pounds of shelled grain.

Subtracting these 1,655 pounds from the 5,845 that theoretically remain after members of the household have been fed, one finds that households have some 4,190 pounds of grain unaccounted for. The answer as to how this grain is used seems to lie with the sizeable domestic animal population. Mrs. Cowgill has suggested, for example, that in calculating family maize consumption, one should always include a dog. For the community she studied in the Petén, she states that families average five persons plus their dog. On this basis, she considers them to be composed, on the average, of six members (U. Cowgill, 1962: 277).

From the size and state of canines in Chichipate, one can hardly argue that they consume as much per capita as humans. However, dogs are only one of several common domestic animals in the community. The animal and fowl population per household in 1965 was, on the average, 11.8 chickens, 4.8 hogs and pigs, 2.3 turkeys, 0.5 ducks, and 0.8 dogs. Except for dogs, all of these are kept by milperos for both family consumption and sale. All must be regularly fed. Swine, particularly, consume enormous quantities of food; a mature hog can eat up to three times as much as can a human.

The remainder of 4,190 pounds of shelled maize hardly seems unaccounted for, then, once the needs of these domestic animals are taken into consideration. In fact, if anything, this remainder seems too small.[5]

To summarize, on the basis of estimates given in thirty interviews, average maize production and consumption per household is as follows:

	6,575	pounds shelled corn, wet season milpa
Plus	1,723	pounds shelled corn, dry season milpa
	8,298	pounds shelled corn per year, total production
Plus	428	pounds shelled corn purchased
	8,726	total available pounds of shelled corn
Less	2,453	pounds consumed by household members
	6,273	pounds left after family consumption
Less	2,083	pounds shelled corn sold
	4,190	pounds available for animal consumption

These figures help in assessing the potential of an agricultural system like that of Chichipate. Certainly, as it stands now, it seems capable of supporting the existing population as well as of exporting considerable quantities of foodstuffs. If the average difference between sale and purchase of shelled maize is 1,655 pounds per year per household, this would amount to an annual export of 148,950 pounds (or some 70+ tons) for the community as a whole, given the fact that in 1965 there were ninety households. If anything, rather than being inflated, this figure seems to be a conservative estimate. Some milperos, for example, pride themselves on selling more than 9,000 pounds of grain after each wet season harvest.

On the basis of her Petén study, Ursula Cowgill argues that, with some help from his wife and children, a milpero can theoretically take care of up to 13 acres and thus provide enough maize for 12.6 people (U. Cowgill, 1962: 277). The Chichipate data indicate that, if one were to count only that maize which family members directly consume and disregard that which is fed to animals and sold, enough is being produced already to provide for the needs of 12.8 people. One could argue that, because this is presently being

5. It is possible that ethnographic reports have consistently underestimated the amount of grain consumed by animals. Compare, for example, the figures here cited with Benedict and Steggerda's impressionistic statement that maize in Yucatán is given chiefly to chickens and dogs in amounts equaling about half as much as the family consumes (Benedict and Steggerda, 1937: 173).

done with only 8.71 acres of milpa, the Chichipate system is more efficient than that described by Cowgill. However, such an argument would be pointless. First of all, our materials are still too much the product of educated guesswork to allow us to make meaningful, precise comparisons. Second, most of the differential seems to arise from the fact that Chichipateños report a lower direct consumption of maize than do Mrs. Cowgill's Petén inhabitants. The fact that they have a ready supply of animal products may be directly related to their lower reported maize consumption, but whether such actually is the case is difficult to say, since Mrs. Cowgill has given us so little information on domestic animals in the Petén.

It does seem evident, however, that the Chichipate data fail to confirm the major point Reina has made in his "Milpas and Milperos" (Reina, 1967: 14). There he argues that milperos in San José suffer a net deficit between production and consumption of 1,300 pounds per family per year. Unfortunately, Reina never tells us what the people do to make up for this deficit, nor does he seriously take into account the production of any foodstuffs other than wet season maize. In addition, in working through his calculations, his almost exclusive reliance on acreage data filed with the municipal government places his whole argument in question. The secretariat of the municipality of El Estor, the body which has been directly supervising land use in Chichipate, has figures on milpa size for the community. The secretary himself, however, considers these figures to be very poor representations of reality. Milperos are reluctant to reveal the extent of their holdings when they know that this information will be used by a government agency; taxes and tribute are real threats in a society as complex as that of modern Guatemala.

Adjusting labor figures proportionately for the type of sites used in 1965, it would seem that milperos spent, on the average, 692 hours per *manzana* in all activities related to wet season cropping that year. Multiplying this figure by that for average milpa size (3.1 *manzanas*), one arrives at an average global input figure of 2,145.2 hours per household. If this figure is then divided by the total number of weeks in which wet season cropping activities, including selection and preparation of site, likely occur (January through October), the average weekly labor input per household comes to some 48 hours and 45 minutes. It must be remembered, however, that such a figure does not represent the labor of only the household head; it also includes that of his sons, sons-in-law, and other helpers. In any given week a household head may himself work only twenty hours or less on his land.

Input of milpa labor is, of course, sporadic. Certain times during the wet season cropping cycle demand extraordinary expenditures of effort; others require practically none. However, the swing between great and little or no activity does not neatly follow the breakdown between the rainy and the dry season, as Drucker and Heizer would have us believe (Drucker and Heizer, 1960: 219); clearing activities make the end of the dry season the busiest time of the year.

Because it requires only sporadic input of labor, slash and burn agriculture, as practiced by Chichipateños, allows for the periodic diversion of energies into other activities. There is a growing consensus that a system such as this could have provided more than sufficient labor to account for the monumental building found in certain prehistoric Maya sites, indeed that even the most elaborate of these structures could have been built with the efforts of a scattered population that came together periodically for a short period of time (Reina, 1967: 18). Charles Erasmus has argued, for example, that even the Uxmal Nunnery, one of the most impressive of the Yucatecan structures, could have been built in thirteen and one-half years if there were a population density of 30 per square mile in the surrounding area, and if each household were to dedicate only 40 man-days per year for the undertaking (Erasmus, 1965: 296).

Today labor left over after agricultural demands have been met is being channeled into the development of neighboring estates and into the occasional need for the maintenance of such public services as roads. Elaborate temple complexes of the type built in the Petén during Classic times neither exist nor are they in the plans. In their stead stand rapidly expanding commercial and administrative centers like Panzós and El Estor. Their existence indicates that lowland slash and burn agriculturalists are on at least as high a level of "socio-cultural integration" as they have been at any time in the past (cf. Erasmus, 1965: 279).

There is no doubt that the Maya lowlands support far fewer people than they could today, even on the basis of the type of slash and burn agriculture practiced in Chichipate. What the maximum carrying capacity of the land might be, given this system, is nevertheless difficult to say. Estimates for milpa-supported population maximums vary from 48 persons per square mile for La Venta Island (Drucker, 1961: 69) to as high as 200 for the Petén (U.

Cowgill, 1961: 53). For Chichipate, given the present slash and burn system and the current length of fallow, the heaviest population density possible would seem to be about 106 persons per square mile of *arable* land. The accompanying land-use projection, based on aerial photographs taken in 1964, will give some indica-

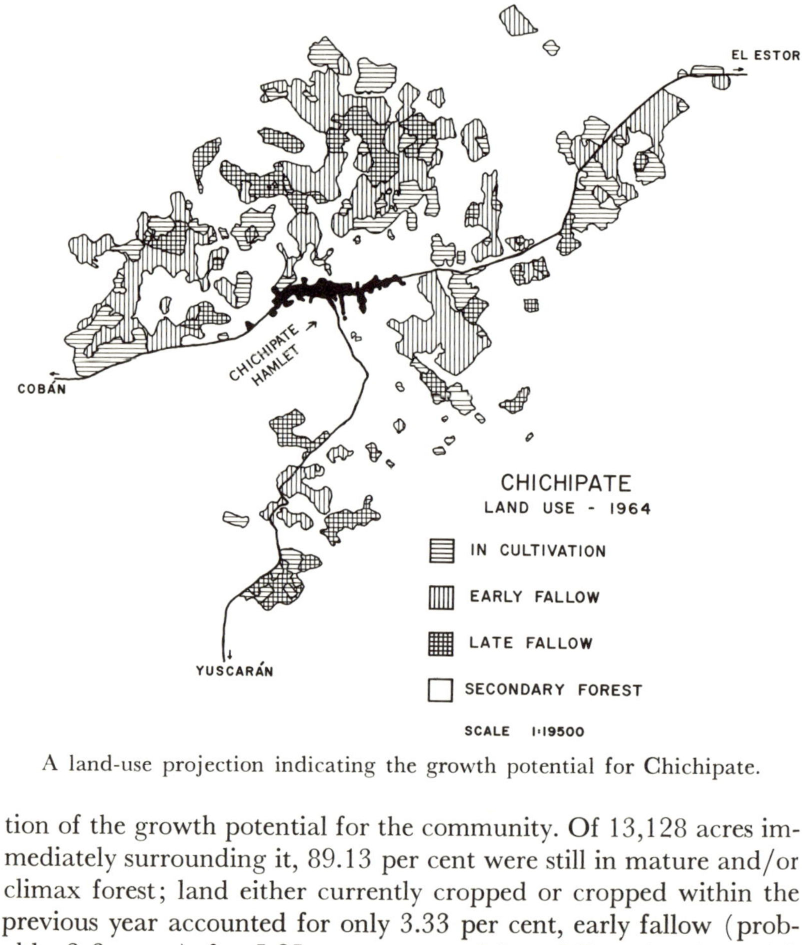

A land-use projection indicating the growth potential for Chichipate.

tion of the growth potential for the community. Of 13,128 acres immediately surrounding it, 89.13 per cent were still in mature and/or climax forest; land either currently cropped or cropped within the previous year accounted for only 3.33 per cent, early fallow (probably 3-6 years) for 5.25 per cent, and late fallow (perhaps 6-7 years) for merely 2.29 per cent. Present population density is approximately 21 persons per square mile.

In the absence of a detailed land and drainage survey, it is impossible to estimate the percentage of unused land around Chichipate that might be arable. Even if it were only 50 per cent (a very low estimate), the area would seem to permit a far denser population than presently exists. That the demographic potentiality of any given site depends in large part on the local quality and lay of the land is reflected in the fact that a direct correlation exists between these factors and the size of the milpas. In general, where yield is large, the land surface needed to support a family is small, and where yield is small, the land surface is relatively great (Drucker and Heizer, 1960: 216). Other considerations, however, give even generalizations such as these a fluctuating rather than static character. Urrutia, for example, has demonstrated that farmers in Uaxactún, by merely reducing plant spacing, can increase yields by as much as 36 per cent (Urrutia, 1967: 91).

The current trend in studies of slash and burn systems in the lowland tropics is to stress their productive capacity. They do have limitations, however. In the virgin rain forest most plant nutrients are tied up in vegetation or rapidly decaying leaf litter, and so evaluations of the inherent fertility of soils that are based on appearance of vegetation type are often misleading. Popenoe has observed that "in many places, almost identical stands of forest trees will be found on fertile and unfertile soils lying side by side" (Popenoe, 1960: 131). Population pressure can quickly bring about total impoverishment of the poorer soils and result in rapid damage by erosion (Popenoe, 1960: 139).

Some of the most highly developed slash and burn systems today are located in West Africa. There, under the stimulus of a sympathetic sociopolitical organization, an impressive amount of foodstuffs is produced for export to the cities and abroad (Bronson, 1966: 272). The major ingredient they contain, that seems to be lacking in the Maya lowlands of Guatemala, appears to be this sympathy from the sociopolitical organization of the nation-state. The problem, in a sense, may be reduced to the dual factors considered so essential by early Classic economists: land and labor. Each is controlled, respectively, by persons owing primary allegiance to different cultural systems. The highland Indians seem more than willing to go to the lowlands if, by doing so, they can be their own masters and can have an outlet for their products. Many landowners, for their part, are more than anxious to begin to develop their holdings, but they suffer from shortages of capital and labor.

Roads are being opened up in the lowlands today, even in the middle of the Petén (Adams, 1965). Thus, the problem of accessibility seems on the way to solution. The problem of control over basic productive resources continues, however. The Indian hoards labor and is hesitant to loosen it to develop the lowlands, except on his terms and with his own techniques. The absentee landowner hoards land. He is generally appalled by what he considers to be the destructive techniques of Indian agriculture and would like to see the lowlands developed on the basis of large-scale, commercialized enterprise. While he may tolerate a few slash and burn squatters, seeing in them a source of needed labor, he certainly does not want all his land taken over by them.

An idea as to the magnitude of this land access problem may be gained by looking at tenure figures for the municipality of El Estor.[6] For the eighty officially registered separate parcels of land in the municipality, the size ranges from 111 to 4,448 acres, with the mean resting at 2,649 and the median at 3,336 acres. Most of these parcels are either uninhabited or else very sparsely populated today. Some landowners have not even seen their holdings. Others have grouped a number of separately registered parcels and so control tens of thousands of acres.

Cases like that of Chichipate indicate that highland Indian groups can successfully move to the lowlands and, with a minimum of adjustment, work out a productive agricultural system. This is possible mainly because of two things: (1) maize is the principal staple crop both in highlands and lowlands, and (2) slash and burn as practiced by these Guatemalans is fairly simple in comparison with other slash and burn systems. Plant cover is classified more by size and density of growth than by specific associations or particular species; basic cultigens are few; soils are typed according to the simple categories of color, drainage, and texture; and common tools consist of merely an axe, a machete, a dibble, a digging stick, and fire. With a minimum of technological adjustment, then, a milpero can go from highlands to lowlands and back again. Never does he have to undertake the type of major systemic change that cultivators from the high Andes must face when they move into the Amazon lowlands. And this perhaps explains why pre-Conquest high culture flourished in both highland and lowland Guatemala, whereas, in Peru, it could never successfully establish itself east of the Andes.

The fact that Guatemalan highland farmers can cope so successfully with conditions in the lowlands does not solve the problem of

6. These come from the most recent source available. Regrettably, it is now over twenty years old (cf. *Indice,* 1944).

lowland development, however. In assessing the potentialities of their northern lowlands, Guatemalan administrators are faced with a very complex dilemma. They can confiscate unused or only slightly used private lands and offer these, together with previously held government lands, for spontaneous colonization by slash and burn milperos. If, as they do this, they sponsor road-building projects to provide these lands with ready access, the lowlands will probably be settled and made to produce very rapidly. In following such a course of action, however, the government would face at least two major problems. First, it would be betraying its commitment to a basic component of its present economic system—the right of private landholding; it would weaken investors' confidence in the country even more than it is weakened at the present time. Second, while it would be opening an escape valve for the population pressure of certain highland departments and would be stimulating an increase in food production, it would also be limiting future population density in the lowlands to that sustainable by an extensive slash and burn system. Once milperos overran the region on their own terms, it would probably be difficult to get them to turn to commercialized or intensive agriculture.

Another alternative open to the government is to encourage absentee landowners to capitalize and develop their holdings and to break up present government holdings among other likely outside investors. Here the basic problems are that many of the present landowners lack sufficient capital and, even when they have it, they find themselves stymied because of the absence of willing labor. The Guatemalan government could wait until increasing population pressure forces highland laboring groups to offer their skills to lowland landowners, whether they want to or not. But this development would lie far in the future and would involve considerable human suffering at both ends of the migrant continuum.

The situation is far more complex than certain writings in cultural ecology would seem to acknowledge (e.g., Steward, 1955). The present level(s) of technology in Guatemala, when applied to the northern lowlands, permits several possible developmental states and concomitant social arrangements. The crucial factors intervening between the natural environment and the social structure involve more than mere technology. Absentee Guatemalan landlords prize private ownership of large units of land; even if these units lie idle and unproductive, they symbolize well-being and the hope for a better future. For their part, highland Indians prize freedom of movement in rotating crops and in disposing of the fruits of these crops. Truly empty lands lure them because of their promise of such freedom; lands in the hands of absentee owners do not. The conflict between the two major cultural systems that are found today in the northern lowlands hinges largely on the question of meaningful goals, i.e., values. Until a better compromise can be worked out between these two systems, a continuation of the present developmental stalemate seems inevitable.

Epilogue

THE FIRST DRAFT OF THIS MONOGRAPH was written before I returned to Chichipate in August, 1967. The few days spent there at that time convinced me that Chichipate, as an agricultural hamlet, is probably doomed. Whether it survives in another form will depend on many factors.

One alternative stimulus for opening up the lowlands mentioned in the previous chapter is the importation of foreign capital. I expressed skepticism that this could ever reach major importance for the area. In the case of Chichipate, however, it has succeeded in doing so. The International Nickel Company has inflicted major alterations on the landscape in the last two years. In preparation for shipping ore from what is reported to be one of the richest nickel deposits in the world, the company has planned the construction of docks just outside of El Estor. Vast expansions of bush have been denuded in anticipation of the mining itself. In Chichipate, milperos have been hired to clear a jagged spur of limestone that cuts into the heart of the hamlet. Here plans call for the immediate building of a cement factory.

Chichipateños are delighted over the course of events. The mining company, in 1967, was paying 1.40 quetzales per day for common laborers; in comparison with the 50 cents workers previously earned, this seemed like a fortune. The owner of Yuscarán, in order to continue recruiting labor, has been forced to up his pay to 75 cents per day, even for those who are squatting on his land. Rumor has it that he is so consistently losing money as to think of pulling out entirely from the venture.

News of the nickel company's operations travels fast and brings migrants from all over Guatemala. The population of the town of El Estor grew from 1,632 in 1965, to more than 2,500 in 1967. Chichipate itself expanded from 90 households, accounting for 432 people, to 126 households, accounting for 604 people. Both the nickel company and the owner of Yuscarán have tried to prevent this expansion, but have been unsuccessful. Squatters have overrun lands that were off limits in 1965, and in 1967 the absentee owners appeared helpless to remedy the situation. Chichipateños themselves are bragging that now they can cut mature forest wherever they wish. The only catch is that now there remains relatively little mature forest land within easy walking distance of the hamlet. There is no more loaning and borrowing of land; each either has his own or has none. As a sort of last ditch effort, in 1967, a representative of the mining company was coming into the community every eight days with orders that no other squatters be admitted.

Chichipate is being swept into the modern world of international mining neither because of, nor in spite of, its milperos. In their desire for freedom of action, in their search for rich new lands, in their rapid adjustment to moist, hot, tropical conditions, Chichipateños still stand as an important example of Guatemala in change. They teach us that, even with the simplest of slash and burn methods, the lowlands hold considerable potential for development. Whether that potential will be tapped as they have tapped it, or in some other way not yet clearly seen, only the future can say.

References Cited

Adams, Richard N. 1965. Migraciones Internas en Guatemala: Expansión agraria de los indígenas Kekchíes hacia el Petén. Estudios Centroamericanos No. 1. Guatemala, Seminario de Integración Social.

Allan, William. 1965. The African Husbandman. Oliver and Boyd, Edinburgh.

Barrau, Jacques. 1958. Subsistence agriculture in Melanesia. Bernice P. Bishop Museum Bulletin 219, Honolulu.

———. 1959. Bush fallowing system of cultivation in the continental islands of Melanesia. Bangkok, Proc. Pacific Sci. Congr. Pacific Sci. Assoc., 9th, 1957, 7: 53-55.

Benedict, Francis G. and Morris Steggerda. 1937. The food of the present-day Maya Indians of Yucatán. Washington, Carnegie Institution. *In* Publication No. 456: 55-188.

Borhegyi, Stephan F. 1956. The development of folk and complex cultures in the southern Maya area. Am. Antiquity 21: 343-56.

Bouyoucos, G. J. 1936. Directions for making mechanical analysis of soils by the hydrometer method. Soil Sci. 42: 225-29.

Brainerd, George W. 1954. The Maya Civilization. Southwest Museum, Los Angeles.

Brigham, William T. 1887. Guatemala, the Land of the Quetzal. Charles Scribner's Sons, New York.

Bronson, Bennet. 1966. Roots and the subsistence of the ancient Maya. Southwest. J. Anthropol. 22: 251-79.

Brown, Paula and H. C. Brookfield. 1959. Chimbu land and society. Oceania 30(1): 1-75.

Budowski, Gerardo. 1956. Tropical savannas, a sequence of forest felling and repeated burnings. Turrialba 6.1-2: 23-33.

———. 1963. Forest succession in tropical lowlands. Turrialba 13(1): 42-44.

Coe, Michael D. 1965. A model of ancient community structure in the Maya lowlands. Southwest. J. Anthropol. 21: 97-114.

Conklin, Harold C. 1957. Hanunóo agriculture in the Philippines. F.A.O. Forestry Development Paper No. 1. Rome, Food and Agriculture Organization of the United Nations.

———. 1961. The study of shifting cultivation. Current Anthropology 2(1): 27-61.

Cook, O. F. 1921. Milpa agriculture, a primitive tropical system. Washington, Ann. Rept. of the Smithsonian Institution for 1919: 307-26.

Cowgill, George L. 1963. Postclassic Period Culture in the Vicinity of Flores, Petén, Guatemala. Unpublished Ph.D. dissertation. Harvard University, Cambridge.

———. 1964. The end of classic Maya culture: A review of recent evidence. Southwest. J. of Anthropol. 20: 145-59.

Cowgill, Ursula M. 1960. Soil fertility, populations and the ancient Maya. Proc. Natl. Acad. Sci. U.S. 46: 1009-11.

———. 1961. Soil fertility and the ancient Maya. Trans. Conn. Acad. Arts Sci. 42: 1-56.

———. 1962. An agricultural study of the southern Maya lowlands. Am. Anthropol. 64: 273-86.

Cowgill, Ursula M. and G. E. Hutchinson. 1963a. El Bajo de Santa Fe. Trans. Am. Phil. Soc. 53: 1-51.

———. 1963b. Ecological and geochemical archaeology in the southern Maya lowlands. Southwest. J. Anthropol. 19: 267-86.

———. 1963c. Sex-ratio in childhood and the depopulation of the Petén, Guatemala. Human Biol. 35: 90-103.

Denevan, W. M. 1961. The upland pine forests of Nicaragua: A study in cultural plant geography. Univ. Calif. (Berkeley) Publ. Geog. 12: 251-320.

Departamento de Censos y Encuestas. 1966. Censo de población 1964: resultados de tabulación por muestreo. Guatemala, Dirección general de estadística.

Drucker, Philip. 1961. The La Venta Olmec support area. Kroeber Anthropological Society Papers 25: 59-72.

Drucker, Philip and Robert F. Heizer. 1960. A study of the milpa system of La Venta Island and its archaeological implications. Southwest. J. Anthropol. 16: 36-45.

Duellman, William E. 1963. Amphibians and reptiles of the rainforests of southern El Petén, Guatemala. Univ. of Kansas Publications, Museum of Natural History 15(5): 205-49.

Dumond, D. E. 1961. Swidden agriculture and the rise of the Maya civilization. Southwest. J. Anthropol. 17: 301-16.

Dunn, Emmett Reid. 1944. Los generos de amfibios y reptiles de Colombia, III. Bol. Inst. Cienc. Nat. Univ. Nacional Colombia 3: 155-224.

Emerson, Robert A. n.d. A Preliminary Survey of the Milpa Systems of

Maize Culture as Practised by the Maya Indians of the Northern Part of the Yucatán Peninsula. Cambridge, mimeographed 14 pages.

Erasmus, Charles J. 1965. Monument building: Some field experiments. Southwest. J. Anthropol. 21: 277-301.

Ferdon, E. H. 1959. Agricultural potential and the development of cultures. Southwest. J. Anthropol. 15: 1-19.

Foster, G. M. 1953. What is folk culture? Am. Anthropol. 55(2): 159-71.

Gann, T. W. F. 1918. The Maya Indians of southern Yucatán and northern British Honduras. Washington, Bureau of American Ethnology Bulletin 64.

Geertz, Clifford. 1963. Agricultural Involution. University of California Press, Berkeley.

Harris, Marvin. 1964. Patterns of Race in the Americas. Walker, New York.

Haviland, William A. 1966. Social integration and the classic Maya. Am. Antiquity 31: 625-31.

Heizer, R. F. 1960. Agriculture and the theocratic state in lowland southeast Mexico. Am. Antiquity 26: 215-22.

Hester, Joseph A., Jr. 1954. Natural and Cultural Bases of Ancient Maya Subsistence Economy. Doctoral dissertation, University of California at Los Angeles.

Higbee, E. 1948. Agriculture in the Maya homeland. Geograph. Rev. 38: 457-64.

Indice. 1944. Indice de los Expedientes que corresponden al archivo de la escribanía y sección de tierras . . . hasta julio de 1943. Guatemala, Tipografía Nacional.

Kaufmann, Terrence S. 1964. Materiales linguísticos para el estudio de las relaciones internas y externas de la familia de idiomas Mayanos. *In* Desarrollo Cultural de los Mayos. Mexico, D.F., Universidad Autónoma de Mexico.

Kelly, Isabel and Angel Palerm. 1950. The Tajín Totonac. Washington, Smithsonian Institution, Institute of Social Anthropology, Publication No. 13.

Kempton, James H. n.d. Preliminary Report of the Agricultural Survey of Yucatán of 1935. Cambridge, mimeographed 10 pages.

Landa, Diego de. 1938. Relación de las Cosas de Yucatán. Introducción y notas por Héctor Pérez Martínez. 7 ed., con un apéndice en el cual se publican por primera vez varios documentos importantes y cartas del autor. Mexico, Editorial Pedro Robredo.

Leach, E. R. 1959. Some economic advantages of shifting cultivation. Bangkok, Proc. Pacific Sci. Congr. Pacific Sci. Assoc., 9th, 1957, 7: 64-66.

Lundell, Cyrus L. 1933. The agriculture of the Maya. Southwest Review 19: 65-77.

———. 1937. The vegetation of Petén. Washington, Carnegie Institution Publication No. 478.

MacKie, Euan W. 1961. New light on the end of Classic Maya culture at Benque Viejo, British Honduras. Am. Antiquity 27: 211-14.

McQuown, N. 1964. Los origines y la diferenciación de los Mayos según se infiere del estudio comparativo de las lenguas Mayanas. *In* Desarrollo Cultural de los Mayos. México, D.F., Universidad Autónoma de México.

Morgan, W. B. 1957. Some comments on shifting cultivation in Africa. Research Notes. Department of Geography, University College of Ibadan 9: 1-10.

Morley, Sylvanus G. 1947. The Ancient Maya. 2nd ed. Stanford University Press, Stanford, California.

Poleman, Thomas T. 1964. The Papalopan Project. Stanford University Press, Stanford, California.

Popenoe, Hugh. 1959. The influence of the shifting cultivation cycle on soil properties in Central America. Bangkok, Proc. Pacific Sci. Congr. Pacific Sci. Assoc., 9th, 1957, 7: 72-77.

———. 1960. Effects of Shifting Cultivation on Natural Soil Constituents in Central America. Doctoral dissertation. University of Florida, Gainesville.

Redfield, Robert and Alfonso Villa Rojas. 1962. Chan Kom, a Maya Village. University of Chicago Press, Chicago, Ill.

Reina, Ruben E. 1963. The abandonment of primicias by Itza of San José, Guatemala, and Soccotz, British Honduras. Tikal Report No. 10.

———. 1964. The urban world view of a tropical forest community in the absence of a city, Petén, Guatemala. Human Organization 23: 265-77.

———. 1966. The Law of the Saints: A Pokoman Pueblo and its Community Culture. Bobbs-Merrill Publishing Co., Indianapolis, Ind.

———. 1967. Milpas and milperos. Am. Anthropol. 69(1): 1-20.

Sanders, W. T. 1962. Cultural ecology of the Maya Lowlands. Estud. Cultura Maya 2: 79 ff.

Sauer, Carl. 1958. Man in the ecology of tropical America. Bangkok, Proc. Pacific Sci. Congr. Pacific Sci. Assoc., 9th, 1957, 7: 105-10.

Simmons, Charles Shaffer, J. M. Tárano, and J. H. Pinto. 1959. Clasificación de reconocimiento de los suelos de la República de Guatemala. Guatemala, Editorial del Ministerio de Educación Pública.

Smith, Hobart M. and E. H. Taylor. 1945. An annotated checklist and key to the snakes of Mexico. Smithsonian Institution, U.S. National Museum Bulletin 187.

Soustelle, J. 1937. La Culture Matérielle des Indiens Lacandones. Paris, Société des Américanistes.

Souza-Novelo, Narciso. 1950. Plantas alimenticias y plantas de condimento que viven en Yucatán. Mérida, Yucatán, Instituto Técnico Agrícola Henequenero.

Stadelman, Raymond. 1940. Maize cultivation in northwestern Guatemala. Contributions to American Anthropology and History. Washington, Carnegie Institution 6(33): 83-266.

Steggerda, Morris. 1941. Maya Indians of Yucatán. Washington, Carnegie Institution Publication 531.

Stevens, Rayfred. 1964. The soils of middle America. *In* R. West [ed.], Handbook of Middle American Indians. Vol. 1. University of Texas Press, Austin.

Steward, Julian H. 1955. Theory of Culture Change. University of Illinois Press, Urbana.

Stuart, L. C. 1964. Fauna of middle America. *In* R. West [ed.], Handbook of Middle American Indians. Vol. 1. University of Texas Press, Austin.

Thompson, J. Eric S. 1927. The civilization of the Mayas. Chicago, Chicago Natural History Museum Popular Series, Anthropology 25.

———. 1930. Ethnology of the Mayas of southern and central British Honduras. Chicago, Field Museum of Natural History, Publication No. 274, Anthropological Series Vol. XVII, No. 2.

———. 1938. Sixteenth and seventeenth century reports of the Chol Mayas. Am. Anthropol. 40: 584-604.

Tondeur, G. 1955. Shifting cultivation in the Belgian Congo. Unasylva 9(2): 67-71.

Urrutia, Victor M., Jr. 1967. Corn Production and Soil Fertility Changes under Shifting Cultivation in Uaxactún, Guatemala. Master's thesis. University of Florida, Gainesville.

Vogt, Evon Z. 1961. Some aspects of Zincantan settlement patterns and ceremonial organization. Estud. Cultura Maya 1: 131-45.

Wisdom, C. 1940. The Chorti Indians of Guatemala. University of Chicago Press, Chicago, Ill.

Wolf, E. R. 1959. Sons of the Shaking Earth. University of Chicago Press, Chicago, Ill.

Wright, A. C. S., D. H. Rommey, R. H. Arbuckle, and V. E. Vail. 1959. Land in British Honduras. Report of the British Honduras Land Use Survey Team. London: Her Majesty's Stationery Office. Colonial Research Publication No. 24.

LATIN AMERICAN MONOGRAPHS—SECOND SERIES

NUMBER 1 (1965): *Fidel Castro's Political Programs from Reformism to "Marxism-Leninism"*
by Loree Wilkerson

NUMBER 2 (1966): *Highways into the Upper Amazon Basin*
by Edmund Eduard Hegen

NUMBER 3 (1967): *The Government Executive of Modern Peru*
by Jack W. Hopkins

NUMBER 4 (1967): *Eduardo Santos and the Good Neighbor, 1938-1942*
by David Bushnell

NUMBER 5 (1968): *Dictatorship and Development*
The Methods of Control in Trujillo's Dominican Republic
by Howard J. Wiarda

NUMBER 6 (1969): *New Lands and Old Traditions: Kekchi Cultivators in the Guatemalan Lowlands*
by William E. Carter